MW01093630

Reckonings

Reckonings

Numerals, Cognition, and History

Stephen Chrisomalis

The MIT Press
Cambridge, Massachusetts
London, England

This book was set in Stone Serif and Stone Sans by Westchester Publishing Services. Printed and bound in the United States of America.

Library of Congress Cataloging-in-Publication Data is available.

ISBN: 978-0-262-04463-9

10 9 8 7 6 5 4 3 2 1

Contents

Acknowledgments

This book has been over a decade in its conceptualization. It began around the time of the publication of my first monograph, *Numerical Notation: A Comparative History*, in 2010, which is an encyclopedic, dense piece of writing intended for a specialist audience. Discussions with my colleagues suggested that a book that better conveyed both the joy and importance of the subject of numeral systems might interest a broader range of readers, and the talks and essays that grew into this book are a product of those conversations. My thanks to all of you, whether named below or not, for your encouragement throughout this process.

Chapter 1 is a significantly revised version of an article, "Constraint, Cognition, and Written Numeration," which appeared in *Pragmatics and Cognition* 21 (3) (2013): 552–572, and appears here courtesy of John Benjamins Publishing Company (Chrisomalis 2013). Earlier versions were presented as talks at Wayne State University. Many thanks to Jérôme Rousseau, André Costopoulos, Giovanni Bennardo, Todd Meyers, David Olson, Michael Thomas, Summar Saad, and Tyler Brockett for their useful comments on various drafts.

Chapter 2 and chapter 5 began life, in very different ways, as a talk at the Language, Culture, and History conference organized at the University of Wyoming, and later in a heavily revised version as a talk at the Program in World Philology sponsored by the Heyman Center for the Humanities at Columbia University. Many thanks to Leila Monaghan, David Lurie, Sheldon Pollock, the late Michael Silverstein, and the late Bernard Bate for their comments and thoughts on these talks.

Chapter 3 is an expanded and heavily revised version of talks given at the University of Bergen SPIRE workshop, "The Cultural Dimensions of Numerical Cognition." Many thanks to Andrea Bender, Marie Coppola, Fiona Jordan, Geoff Saxe, Rafael Núñez, Karenleigh Overmann, Dirk Schlimm, and the late Sieghard Beller for their comments. John Bodel, John Dagenais, and Paul Keyser provided additional useful evidence and reflections necessary for the final version of the essay.

Chapter 4 had its genesis in talks given at the Society for Anthropological Sciences annual meetings, while talks at the International Medieval Congress in Kalamazoo, Michigan further expanded these ideas and provided evidentiary support for my suppositions. Thanks to Amalia Gnanadesikan, Shana Worthen, Tom Abel, Jim Plamondon, and David Kronenfeld for conversations and remarks on this work.

Chapter 6 is the product of ideas first developed as part of the "Signs of Writing" workshop series organized by Chris Woods and Ed Shaughnessy, and later presented at the Oriental Institute of the University of Chicago. Conversations with Chris Woods, Matt Stolper, Andréas Stauder, John Baines, Stephen Houston, Marc Zender, Simon Martin, David Lurie, Pascal Vernus, Nicholas Sims-Williams, and Xiaoli Ouyang have been particularly fruitful at various junctures.

Chapter 7 had its inspiration in early discussions with André Costopoulos leading up to our joint talk "Is Anthropology Ready for Ultimate Questions?" given at McGill University and at Wayne State University's Humanities Center, and later at the American Anthropological Association under the title "Is the Past Like the Present?" It has a complex history and has benefited from discussions with colleagues in anthropology, history, linguistics, and cognitive science. Bill Balée, Doug Hume, Julie Hofmann, Steve Muhlberger, Richard Lee, and many others have been willing to share with me their thoughts on cognitive anthropology, medieval history, and anthropology's "big questions."

My colleagues at Wayne State University remind me every day what a privilege it is to work alongside them. In particular, my thanks to Tamara Bray, Sherylyn Briller, Yuson Jung, Tom Killion, Julie Lesnik, Guérin Montilus, Geoff Nathan, Andy Newman, Ljiljana Progovac, Martha Ratliff, Jess Robbins, Krysta Ryzewski, Andrea Sankar, Susan Villerot, and Margaret Winters. Many thanks also to all my students. Many of the ideas developed here were refined in a graduate seminar, "Anthro-X," several years ago; thanks to Heather Brooks, Sarah Carson, Grace East, Molly Hilton, Summar Saad, and Michael Thomas for their input at a critical juncture.

When I began writing about numerical notations over twenty years ago, there were virtually no typefaces available that encoded the immense variety of the world's numeral systems, requiring me to spend hundreds of hours designing my own. Since that time, through the work of the Unicode Consortium and thousands of scholars and volunteers, it is now possible to encode hundreds of scripts (and number systems) using widely accessible fonts. It has been particularly rewarding to see my own earlier work cited and used in proposals to encode numerical notations in Unicode. The text you are reading exists in part because of the labor of these unheralded participants in a collective effort to ready all scripts and notations for digital use.

My family has been an ongoing source of motivation and encouragement over the past decade. Arthur Chrisomalis continually reminds me of the joys of fatherhood by allowing me to share my passions with (sometimes) interested young people. My father, Peter Chrisomalis, and my father-in-law, Andrew Pope, have been particularly important motivators in producing this book in the way I did—to let me share my ideas widely to broad audiences of people interested in history and mathematics. As always, Julia Pope, my wife, has been a close reader of my text, a careful critic when needed, a passionate advocate in tough times and joyous ones, and as always, an inspiration. My thanks to all.

My thanks to the staff at the MIT Press for their tireless work and thorough professionalism on a complex manuscript. In particular, my thanks to Phil Laughlin, my editor, and Matthew Abbate, my copyeditor, for their attention to detail and their flexibility throughout this process.

This book would never have existed without the care and mentorship of my doctoral supervisor, the late Bruce Trigger (1937–2006) at McGill University. Although not one word in this book was written during his lifetime, every single word—and particularly the last paragraph of the text—bears the imprint of his work with me. This book is dedicated to Bruce in honor of his decades of commitment both to his students and to comparative cross-cultural research in anthropology and archaeology.

Introduction: Three reckonings

This is a book of three reckonings.

Firstly, to reckon is to calculate. In the modern world, calculating, working, and playing with numbers are far more important activities than in any earlier period. However, numeracy (the quantitative counterpart of literacy) is fundamental to the functioning of all large-scale societies, including premodern ones. It is central to trade, the appropriation of wealth, and the redistribution of surpluses. And working with numbers, at a smaller scale, is part of all sorts of everyday practices: measuring out quantities for recipes, working out the days until the next season of a favorite show is released, or tallying the number of guests at a party. Reckoning is a (nearly) universal social practice that has taken on new meanings and new functions in capitalist and industrialized societies. The mathematician John Allen Paulos (1988) correctly diagnoses, though, that despite the ubiquitous nature of arithmetical problems in modern, industrial lives, people don't always have the tools needed to work adequately with numbers. Sometimes this can create enormous feelings of inadequacy. As a psychological response, some people take a perverse pride in their innumeracy, in a way that we would never do with illiteracy. In contrast, I admit to taking some perverse interest in the historical and social scientific study of mathematics and numbers, a subject that many of my colleagues who are anthropologists find baffling or even disgusting. (Some of them even got into the discipline to avoid taking more math!)

Reckoning—counting and calculating—keeps a suitable semantic distance from words like *mathematics.* This book is not a history of mathematics, and you do not need any training in the field to read it. Understanding how people, both past and present, reckon—how they work with numbers as part of daily life—helps us grapple with their lifeways. Those of us who learned arithmetic using pen and paper, working with the ten digits 0–9 and place value, may find it natural and take for granted that this is the way it's always been done, or at least the way it ought to be done. But think of the amount of time and energy spent in the early school years just to teach place value, and you'll realize that this sort of numeracy is not preordained. And that's not even considering that pen-and-paper arithmetic requires the widespread availability of cheap writing instruments and media as seemingly simple as paper, which shouldn't be taken for granted.

Over the past 5,500 years, more than 100 structurally distinct ways of writing numbers have been developed and used by numerate societies (Chrisomalis 2010). Thousands more ways of speaking numbers, manipulating physical objects, and using human bodies to enumerate are known to exist, or to have existed. Each of us is familiar with only a tiny fraction of the diversity of the world's number systems. The current universality of a particular set of these practices among Western, Educated, Industrial, Rich, and Democratic (or WEIRD) societies leaves much of the human condition uninvestigated (Henrich, Heine, and Norenzayan 2010). Because this diversity in reckoning practices has only partially been described, we need a better understanding of how people work with numbers. In this book, I draw on, and expand upon, the enormous cross-cultural and comparative literatures in linguistics, cognitive anthropology, and the history of science that bear on questions of numeracy.

Secondly, to reckon is to think. The etymological and semantic linkage between thinking and calculation is strong in English, as it is throughout large swaths of the Indo-European family of languages. We draw semantically on the taxing mental activity of computation, extending it metaphorically to thinking in general. A reckoning is an estimation or judgment that is undergirded by the mental work necessary to reach it. Explaining the history of number systems relies on understanding the mental, verbal, and symbolic manipulations that mathematical cognition requires. These cognitive operations are embodied through language and gesture, and they are materialized through artifacts and notations (Overmann 2016). Because numeracy is not just a social process but also a cognitive one, this is a book about reckoning about reckoning.

This is not a formal cognitive analysis, however. I present no technical accounting of how past people thought while manipulating numbers. Such an analysis is surely impossible for any period but the present. Where relevant, it draws on experimental cognitive psychological analyses of numerical cognition, some of which are informed by anthropological insights (Rips 2011; Carey 2009). But often, historical and cognitive disciplines have worked in parallel rather than in tandem. The anthropologists, historians, and linguists who work on numeracy and numeration must be aware of these important cognitive scientists. These include my fellow cognitive anthropologists, many of whom have undertaken vital empirical studies of numeracy and mathematics in the daily lives of people (Hutchins 1995; Mukhopadhyay 2004; Marchand 2018). In many cases, however, because this is a book centered on past cognition, I draw much more substantially on the work of cognitively informed humanists such as the anthropologist Jack Goody (1977), the historian of science Geoffrey Lloyd (2007), and the classicist Jocelyn Penny Small (1997). These analyses use fragments of historical, linguistic, and archaeological evidence to examine human cognition across long periods, in light of the impossibility of direct experiment or observation. Within anthropology, cognitive archaeology is the subfield most directly concerned with historical processes of numerical cognition, and I certainly draw on that work as well, where relevant (Morley and Renfrew 2010). I also draw on the considerable and growing body of material on numerical cognition in small-scale societies from cognitive anthropology and cross-cultural psychology, which asks ethnographically informed cognitive questions about the lexical, embodied, and material numeracy of groups traditionally disparaged or ignored in histories of mathematics (Saxe 2012; Everett 2017).

Finally, building on these first two reckonings, to reckon is to evaluate, to assess, and to assign worth. This third reckoning is imbued with conflict and tension—as when we are forced to reckon with some new circumstance. We make choices and judgments that have consequences. When it comes to number systems, the prototypical instance of this tension is the replacement of the Roman numerals (for most purposes) by the Indo-Arabic or, better named, the Western numerals. The case study of this replacement forms a thread throughout this book, although it is just one strand of a much broader tapestry. The traditional narrative holds that late medieval and early modern people, evaluating these two systems, chose the more efficient one and abandoned the more cumbersome Roman numerals.

Here I set out a different approach, one that draws on the literature on the disappearance and abandonment of writing systems (Baines, Bennet, and Houston 2011). I acknowledge that people make decisions, and that those decisions have reasons that make sense (both to the decision makers and, with a little work, to us too). If that were not possible, if every "other" were hopelessly foreign to our minds, a cognitive and comparative analysis of numeracy would not be possible. Understanding how people's decision-making processes work, not just at an individual scale but at a collective and social one, helps us understand why one set of practices might not persist. But the reasoning we might use to make those judgments need not be universal—in fact, universal principles of rationality are very unlikely to provide full and satisfactory explanations. Figuring out how and why people abandon one way of doing things in favor of another is not trivial. It is unlikely to be subsumed under a small number of cultural "laws," but it is also too important for us to simply throw up our hands and claim, in a particularistic fashion, that there is no predicting or explaining things.

I recall quite clearly, in the way one recalls times of extreme stress, a question posed to me at my PhD defense in 2003, by the pro-dean of my doctoral committee. This person's job was to read my work from a completely different discipline from mine (in this case, biology), having had no prior contact with me, to ensure that everything was on the up and up, and that they weren't giving away degrees to just anyone. During the defense, this biologist queried whether numerical notations should be conceived as analogues of organisms: after all, they are born, they have their natural life cycle, and then they die. My response at the time, which I still think is right, is that a better analogue is the species. While an organism has a limited and relatively inflexible life course, species evolve (out of ancestral ones), change over time, persist for an indeterminate period, and survive and thrive (or not) in different contexts; when they eventually are replaced or go extinct, it is because they are not well suited for the environment they then find themselves in. Just because one numerical system gives rise to a descendant does not mean that the ancestor must go extinct, and a system that is ideal in one context may be unsuited for another. We must take account of these longer-term, larger-scale historical processes as well as the local and particular.

Of course, the scale of biological evolution and the processes of cultural evolution are quite different, and we must be careful not to ignore the human element in invoking processes like "selection" and "adaptation." One of the most serious and most reasonable charges against cultural evolutionary studies of cultural transmission is that they are insufficiently attentive to the contexts in which decisions to adopt, reject, or transform are made.

So here, in dealing with the reckonings and evaluations of users about which numeral systems to prefer, we must be careful not to assume a universal logic that requires a particular outcome. Because people are not rational choice machines, we need to understand their cultural rationality in order to understand how they made sense of their decisions. This is particularly important when looking at people far removed, chronologically or culturally, from contexts with which we are familiar. Nonetheless, this is an evolutionary book, in a historical sense, seeking to understand the long-term processes by which the reckonings of individuals at discrete periods come together to shape the histories of numerical systems, until they too, inevitably, meet their day of reckoning. Not only are we not at the "end of history" of numeration—surely new systems will be developed and used in the coming centuries—but there can be no such end, as long as humans are still judging and evaluating their numerical tools.

Numerals, cognition, and history: these three reckonings are the three central themes of the following seven chapters. Each element of this triad is essential for understanding human numbering practices and the social context of written numbers. Or so I reckon.

Note: On "Western numerals"

Throughout this book, I use the term "Western numerals" to refer to the set of signs 0123456789, organized in a base-10 system using place value, as I have done in my earlier research. In the English-speaking world, we mostly learn these signs under the name "Arabic numerals," which reflects the fact that they were borrowed by Western Europeans from Arabs living in Spain, Sicily, and North Africa in the tenth century CE. In the scholarly literature on numerals, these are most often called "Hindu-Arabic numerals," which reflects a little more of the history of the system, because the Arabic script got its numerals from an antecedent system used in northern India as early as the fifth or sixth century CE. Other terms like "Indian" and "Indo-Arabic" are also found. The historian of mathematics Carl Boyer, whose early work on numeral systems played an important role in my development as a "numbers guy," argued somewhat facetiously that we might more properly call it the "Babylonian-Egyptian-Greek-Hindu-Arabic" system (1944: 168)—although in this case I think he was wrong, and that "Egyptian-Mauryan-Indo-Arabic" would get the history straight.

The most basic problem with the formulations "Arabic" and "Hindu-Arabic" is that they do not adequately distinguish the set of signs 0123456789 from the set of signs ٠١٢٣٤٥٦٧٨٩ used in Arabic script or the set of signs ०१२३४५६७८९ used in the modern Devanagari script, or any number of other decimal, place value systems. All of these descend ultimately from that same fifth–sixth century CE Indian ancestor. To make matters more confusing, in Arabic the numerals used alongside Arabic script are called *arqam hindiyyah* (Indian numerals). The problem of ambiguity is thus a serious one. Because several such systems are in active use (particularly the Western European 0–9 and the "Arabic" set), it becomes a nightmare to try to distinguish these systems from one another. We need different terms for each set of numerals.

Structurally they are very similar to one another—although not completely; for instance, many Indian writers customarily write 100,000 as 1,00,000 and 1,000,000 as 10,00,000. So I talk about Western, Arabic, and Indian numerals to refer to the decimal, place value systems used in three different script traditions. Paleographically—in terms of the history of the signs themselves—they are quite distinct, and are likely to remain so. One could argue that just as we talk about the "Latin alphabet," we could call 0123456789 the "Latin numerals" instead of "Western." But this would only create confusion with the "Roman numerals." "Western numerals" reflects the fact that the particular graphemes (the specific signs) were developed in a Western European context and were first and most prominently used in Western Europe.

One might argue that by calling them "Western numerals" I am denying them their history, obscuring the fact that they derived from Indian and Arabic notations, which I certainly do not wish to do. But I think that Boyer has a point—why stop at "Hindu," since the "Hindu" place value numerals derive from a nonpositional system used in Brahmi inscriptions in India as early as the fourth century BCE, which in turn probably derives from Egyptian hieratic writing going back as early as the twenty-sixth century BCE! And if we later decide that this history is wrong, do we then change the name? I am far more concerned that by using terms like "Arabic" or "Hindu-Arabic" for 0123456789, we render invisible the continued existence and active use of actual Arabic and Indian numerals in the modern Middle East and South Asia. Using an umbrella term—which, in reality, obscures all but a single variant of a rich family of numerical forms—unfairly collapses this complex genealogy with several extant modern branches into a single unilinear history. The history of place value becomes merely *our* history of place value. And, in the same way that the fallacious evolutionary error that humans are descended from chimpanzees renders chimps as our ancestors when they are actually our cousins, we must avoid rhetoric that suggests that Devanagari, Arabic, Persian, Telugu, Gujarati, and many other decimal positional systems are historical relics. No one seriously disputes the facts of the history and evolution of these systems, but our labeling practices run the risk of making it appear as if we stand alone at the end point of the history of numerals. "Western numerals" highlights that specific paleographic and structural innovations happened in the West (principally in Spain and Italy), but maintains a suitable conceptual distance from the related but still vital systems of the Middle East and South Asia.

1 / I The limits of numerical cognition

> Out of the darkness, Funes' voice went on talking to me. He told me that in 1886 he had invented an original system of numbering and that in a very few days he had gone beyond the twenty-four-thousand mark. He had not written it down, since anything he thought of once would never be lost to him. His first stimulus was, I think, his discomfort at the fact that the famous thirty-three gauchos of Uruguayan history should require two signs and two words, in place of a single word and a single sign. He then applied this absurd principle to the other numbers. In place of seven thousand thirteen, he would say (for example) *Máximo Pérez*; in place of seven thousand fourteen, *The Railroad*; other numbers were *Luis Melián Lafinur, Olimar, sulphur, the reins, the whale, the gas, the caldron, Napoleon, Agustín de Vedia*. In place of five hundred, he would say *nine*. Each word had a particular sign, a kind of mark; the last in the series were very complicated.... I tried to explain to him that this rhapsody of incoherent terms was precisely the opposite of a system of numbers. I told him that saying 365 meant saying three hundreds, six tens, five ones, an analysis which is not found in the "numbers" *The Negro Timoteo* or *meat blanket*. Funes did not understand me or refused to understand me. (Borges 1964: 64–65)

In his short story "Funes the Memorious," Jorge Luis Borges presents an account of a person blessed or cursed with an apparently limitless memory and who, among other things, constructs a bizarre numerical system. Funes's memory allows him to ignore the constraints that apply to those of us whose memories are less prodigious than his own, producing a system of numeral words and number symbols[1] lacking any structure. In so doing, Funes loses the ability to communicate with others or even to perform arithmetic. Borges reminds us that the ability to forget is fundamental to humanity, and that the limitations of human cognitive capacities matter a great deal. Whether we say *seven thousand thirteen* or *Máximo Pérez* is not simply a stylistic choice. Number systems are subject to constraints that limit their range of variability.

In this book, I will use the concept of constraint to ask and answer comparative questions about numerals. Recent cross-cultural work on numerical notation systems helps address longstanding debates in the social and cognitive sciences over the degree to which human behavioral plasticity is limitless. While there is infinite variability in human social life, this does not entail an extreme particularism, nor does it render futile the search for pattern and generalization across time and space. Constraint-based approaches explain probabilistic generalizations about human behavior better than universalistic ones because they allow for exceptions. Actually, such generalizations are best explained by examining and understanding those exceptions.

In examining the interaction of written representations and cognition, there are two general sets of relevant constraints to think about. First, we must ask how perceptual and cognitive factors constrain the structure of numerical notations. In other words, in what ways are the cognitive architecture of humans, and the properties of the immediate environments with which they engage, relevant to the notational systems they employ? This topic has been amply addressed in contemporary Western contexts by cognitive psychologists (Dehaene 2011; Rips 2011). We can look at the properties of notations and see how well they do (or do not) correlate to the properties of the mind relevant to manipulating numbers. This might even let us compare one notation to another in terms of representational efficiency.

Second, we must ask how numerical representations, once invented and used in specific contexts, constrain their users' mathematical cognition. For instance, we might inquire, following the anthropologist Jack Goody (1977: 12–13; 1986: 52–54), how different numerical systems (such as those of the LoDagaa of Ghana or predynastic Sumer) affect arithmetical practices. Goody shows that language and material practice intersect, and while the LoDagaa have a single ordinary number system, they use a special system (*libie pla soro*) for counting cowrie shells in piles of five, twenty, and one hundred, an essential aspect of LoDagaa economic transactions such as bridewealth. As Goody points out, "Counting cows is different from counting cowries"—specifically because the practices involved are different (Goody 1977: 13).[2] Or we might follow the developmental cross-cultural psychologist Geoffrey Saxe (1981, 1982, 2012), whose decades of fieldwork with the Oksapmin of Papua New Guinea show how "body-counting" systems, in which words corresponding to parts of the upper body serve as numerals when pointing to those parts in conversation, respond to changing economic and educational practices. Rather than seeing these systems as vestiges of some imagined prehistory, Saxe argues that counting practices are responsive to changing conditions, and that how users think using them is constrained but not determined by their structure.

Establishing how these two processes intersect—considering cognition both as cause and effect—is critical for a full understanding of how written numeration relates to cognition. Numerical notation has properties that are in some ways similar to those of lexical numeral systems (the number words in any particular language), whose structures are generally regarded by linguists as deriving from universal properties of the human mind. Numerical notation, however, in contrast to language, is not cross-culturally universal and emerges only in specific social contexts. Because it is a graphic representation of number that has no necessary or rigid connection to the number words of any particular language, its properties are different from those of lexical number. That there are two distinct modalities for representing a single domain of activity (number), each with its own structural properties, provides an opportunity to examine the specific role of graphic representation in affecting the structure of cognitive frameworks such as number systems.

Constraints against universals and particulars

By *constraint,* I mean some nondeterministic factor, operating through one or more processes, that makes some outcome more or less likely to occur. A constraint theory helps generate statements about tendencies toward certain outcomes, as well as limitations against some sorts of outcomes, in some area of activity, but it doesn't compel or require a particular outcome. When I was a graduate student I was compelled or required to read endless papers in anthropological and archaeological theory that used phrases such as "lawlike generalizations," which I suppose was meant to imply that some patterns in culture were regular and recurrent but not universal (Watson 1976). However, using such language, at best, serves only as a euphemism for "laws," and at worst conceals a willingness to ignore variation. Constraints are not merely tendencies; a constraint is not simply something that tends (not) to happen, but rather is a factor that helps explain or model that tendency. Additionally, when we focus too much on outcomes, we run into the problem of equifinality, the idea that the same effect may be produced by multiple different causes. In my own discipline of anthropology, the problem of equifinality led Franz Boas (1896) to insist that the study of historical processes—how things came to be the way they were in any particular group—was more important than patterns and structures at any given point in time.

A variety of constraint models have been proposed in the social sciences over the past century. Alexander Goldenweiser's, perhaps the earliest in anthropology, does not use the term "constraint" at all; he speaks instead of the "principle of limited possibilities" (Goldenweiser 1913). This is the very sensible suggestion that for most cultural phenomena, there are not a limitless number, nor only one, but a few distinct configurations. Goldenweiser noted (1942: 124–125), for instance, that despite variability in length, weight, cross-section, material, etc., there are numerous similarities between all workable oars due to physiological, technical, and functional factors—a blade wide enough to propel, a handle thin enough to hold but long enough to serve as a lever, etc. Similarly, you can have a slotted (flat) screwdriver, or a Phillips (cross-head) screwdriver, and many others. In my own home country of Canada, the Robertson (square-head) screwdriver is very common, although rare anywhere else. But you cannot have a workable round-head screwdriver, even though it is easy to imagine one. In fact, I imagine that you are imagining one right now. Even though each context, individual, and history is unique, unrelated groups converge on similar solutions to problems—and some solutions are unworkable, though not unthinkable. Written number systems, similarly, are a notational technology upon which a variety of constraints operate to limit cross-cultural variability.

Goldenweiser's friend Edward Sapir used similar argumentation to explain why, although there are theoretically an infinite number of speech sounds, human languages use only a small fraction of these, because of the cognitive and perceptual limitations of the brain and the muscular limitations of the human mouth and vocal tract (Sapir 1921: 45–52). Thus, some sounds like /m/ are extremely common cross-linguistically, while sounds that are more taxing physiologically, like the two English "th" sounds, are rare. In contemporary linguistics, optimality theory has explicitly used the concept of constraint since its inception (Prince and Smolensky 1993). However, such theories are usually subsumed today under universalizing models that identify absolute and knowable limits on the range of possible languages.

Constraint, as these theories use it, frequently means constraint down to a single solution based on a universal constraint set present in all languages. But constraints on numerical notations cannot be simply derived from universal grammar, because numerical notations are not direct, unmediated representations of language. Just as it matters what the shape of an oar is, it matters what the features of a sign are.

The ecological psychologist J. J. Gibson (1979) developed the concept of *affordances* to describe potentials or possibilities for action that emerge from an object or environmental stimulus. A flat surface affords having something placed upon it, a round object affords being rolled, and fingers afford being counted (and counted upon). An affordance is, in some sense, the opposite of a constraint, which has a limiting effect, but I see no reason to decouple them. Constraints and affordances often co-occur—there is not much difference between a strong constraint against some behaviors and a strong affordance for one particular outcome. Gibson regarded affordances as emerging from the environment (including both *exteroperception*, the perception of the environment, and *proprioception*, the perception of the observer him/herself as part of that environment), and was suspicious of mental models that are not environmentally motivated (Gibson 1979: 141). This reflected his anticognitivist perspective, which was not uncommon at the time he wrote. However, here I part ways with Gibson—cognitive science provides a great deal of support for both constraints and affordances. The affordances come from the environment, but for them to be meaningful for humans, they must be recognized. An affordance-based model may explain why so many languages' numeral words are decimal (because humans have ten fingers) but it does not explain why number words and graphic number signs tend to be ordered from the highest to the lowest powers of the base, for instance.

Here we are playing around with two central theories in social scientific and specifically anthropological approaches. On the one hand, *universalism* is the idea that humans are basically the same, that human behavior differs only to a small and relatively insignificant degree, and thus that findings from one part of the world can be readily extrapolated and generalized. On the other hand, *particularism* argues that the local, historical, and particular conditions of human existence are sufficiently variable that large, generalizing explanations are grossly oversimplifying, and that we would be better to focus ourselves on the panoply of local conditions that cause any particular people to behave the way they do. In his distinguished lecture

"Constraint and Freedom," my doctoral supervisor Bruce Trigger noted that most aspects of human behavior are less orderly than universalism would suggest but more orderly than particularism would suggest (Trigger 1991). Aiming explicitly for a middle ground, he envisioned behavior as the product of a set of overlapping constraints, defined as factors "that human beings must take into account to varying degrees when selecting an appropriate course of action" (1991: 555–556). Trigger distinguished external, universalizing constraints like ecology and technology from internal, particularist ones like symbolic practices and cultural traditions. Trigger's account needs two qualifications. First, it presumes that we are consciously aware of the constraints under which we make decisions; but that is not always the case. Just as individuals do not readily understand the cognitive constraints underlying their phonological processes, most people are unaware of the constraints on their numerical systems. Second, there is no reason to think that internal, mental constraints are less likely to be universal than external, environmental ones. Ecological and environmental variability is enormous, but there are considerable commonalities behind all human brains. The key will be to determine which constraints are relevant, and how.

Constraining infinity

In an old magician's trick, which I first learned about from the popular mathematics of Martin Gardner, the magician asks an audience member to think of a number: "I want you to name a two-digit number between 1 and 50. Both digits must be *odd*, and they must not be alike. For instance, you cannot name 11" (Gardner 1956: 174). After the person has done so, the magician then asks, "Is your number 37?" very frequently to the astonishment of the subject. Partly this trick works because the magician creates rules that restrict the subject's choices, but obscures the severity of the restriction through subtle language that makes it appear as if the subject has more choice. There are actually only eight answers that fulfill these criteria (13, 15, 17, 19, 31, 35, 37, 39). But this cannot be the only explanation, because far more than one-eighth of respondents will respond 37 (otherwise it would not be a very reliable trick). Of course, not everyone will choose 37. That would suggest that there were no other options at all. Similarly, it is uninteresting if somebody chooses to answer 53 or 8 or some other number that violates the specified constraints. The fact that few people choose 15, while lots of people choose 37, is best explained by the hypothesis that three and seven are more prototypically odd than one, five, and nine. Prototype theory, in cognitive science, acknowledges that some members of a category are more central than others, more likely to come to mind rapidly and early (Rosch 1973; Lakoff 1987). So, for most Western readers, *apple* is a prototypical fruit. And, at least in Western societies where 3 and 7 are rich with evocative symbolic meanings, not to mention that they are prime (unlike 1 and 9), and not part of decimalized currency systems like 5, they are also prototypical odd numbers. When some outcome is possible, but is attested significantly more rarely (or more often) than expected, constraints are probably involved.

Now to revise the game somewhat, I might ask you to think of a number that is greater than one but less than two (i.e., neither one nor two themselves). In virtually any sizeable group to whose members such a question would be meaningful, it is a very reasonable bet that some respondents will choose one and a half (1.5). Halves and quarters, and to a lesser degree tenths, are the numerals between the whole numbers that most people use most often. But note that while I did impose a constraint of magnitude, you were free to choose any rational number, like 1.832, or an irrational number, like the square root of 2 (roughly 1.414). There are infinitely many numbers between 1 and 2, and you could have chosen any one of them—far greater than the eight choices you had in the first game. So we have, on the one hand, a sort of constraint, leading most of you to choose familiar numbers within the prescribed limits, but on the other hand, infinite variety—you could have chosen, as I did, 1.6402895036, although you didn't. However, there are also an infinite variety of numbers that you could not have chosen—for instance, 6 or π.

I believe that human cultural variability is analogous to the second game, for many domains of experience. It is often treated as an anthropological axiom that culture is infinitely variable. I do not wish to disagree with this at all. It is self-evident that human cultural variability is not only infinite but increasing over time. Every time two strangers meet, every time someone is born or dies or does anything, something new happens. But "infinity" does not equal "anything can happen." *Cultural variability is infinite, but it is not limitlessly infinite.*

Our task, then, is to figure out what limits exist on different sorts of human behavior. This theory of constraints seeks to understand variation, and then asks what overarching factors or processes limit that variation. This is neither a universalist nor a particularist approach, but a *comparative* one. It does not assume in advance, but rather asks, to what degree and by what processes any aspect of human behavior is constrained. The importance of a theory of constraint is not its moderateness; it is not a Solomonic compromise. Just as you could not have answered 6 within the rules of my second game, there are an infinity of potential human behaviors that are not attested, despite the infinity of those that are. The role of constraint theory is to bridge the gap between the imaginable and the attested, the commonplace and the rare. Without denying infinite variability, it forces us to ask, given considerable cross-cultural variability, what is it that we do not see—and why.

One useful example of this principle comes from Brown and Witkowski's (1981) linguistic and anthropological examination of figurative expressions in 118 languages, which revealed that the figurative association between the pupil of the eye and a small person or child (as in English *pupil* "student") is far more common than would be expected by chance—around one-third of the languages in their sample made this association. The explanation can hardly be genetic, can hardly be random, and does not result chiefly from language contact. Instead, there is a perceptual constraint at play: if one looks into the pupil of someone's eye, one sees a small reflection of oneself (Tagliavini 1949; Brown and Witkowski 1981: 601). While this frequent conceptual linkage invokes general principles that extend beyond any one cultural context, this finding is hardly "universalist," as Brown and Witkowski describe it. A universal that applies to only one-third of the world's languages is hardly a universal, and I do not find their rephrasing of the matter as "universalist tendencies" to be analytically helpful, any more than "lawlike generalizations" avoids the problem of "laws." While Brown (1976) shows that there are universals of body-part nomenclature, they are of the form "All languages have a word for *eye*" rather than these less universal yet more interesting patterns.

Daniel Everett's (2005) controversial account of what he calls "cultural constraint" in the language and behavior of the Pirahã of Brazil is the most startling example of a sort of a theory of constraints in the domain of number. Everett believes the Pirahã to be the only exception to a wide range of otherwise human universals (color terms, embedding), to have the simplest kinship terminology, phonological inventory, and pronominal systems of any language, and to lack number words completely (2005: 622). His explanation for this is decidedly particularistic: the Pirahã possess a unique yet unspoken "immediacy of experience" cultural constraint that restricts their speech and behavior in numerous respects. Everett sees his work as refuting Chomsky's universal grammar, and it is important in expanding our awareness of human variation in the domain of number and many others. Yet it is not clear what exactly he means by "cultural constraint" in this account. If he means that it is a cultural value that is acquired through socialization, then it is not clear how it is acquired, whether it is ever expressed explicitly, whether Pirahã children violate this principle, and what happens if they do. If he means that in acquiring the Pirahã language, individuals are precluded or constrained from certain forms of thought, then this is not a cultural constraint but a linguistic constraint, and becomes circular (language constrains language). What exactly Everett means by "culture" is altogether unclear; accounts of limited-numeral languages generally invoke a social environment in which counting is irrelevant to most decision making, but Everett does not take this functionalist approach. Everett's work has sparked a host of responses and counterarguments, and is still being debated on evidentiary grounds (Frank et al. 2008; Nevins, Pesetsky, and Rodrigues 2009; De Cruz, Neth, and Schlimm 2010; Reboul 2012). However, the chief issue is that there is no explanatory mechanism proposed by which this constraint, and its consequent linguistic expressions, comes into existence and is reproduced and transmitted socially. Without it, Everett's model does not actually provide evidence for a constraint, and he cannot establish to what degree other, similar societies possess similar features (or not), and for what reasons.

Constraining spoken and written numerals

In 1978, Joseph Greenberg published his now justly renowned paper "Generalizations about Numeral Systems" (Greenberg 1978). It is a well-known exposition of a universalist position in linguistics, describing 54 generalizations that Greenberg felt held true of all (or most) number word systems. Greenberg's model proposes that there are strong and unexpected regularities in language structure (particularly syntactic and morphological ones, in this case) that suggest a universal mechanism for their production. While Greenberg presented his rules along with any exceptions he had found, his account is clearly a universalist one in which exceptions are downplayed.

Some of Greenberg's regularities are universal but unsurprising; for instance, that there is no language in the world that has a word for *two* but none for *one*. This generalizes to the regularity that languages with an ordinary word for a natural number $n + 1$ will always have a number word for n.[3] On the other hand, some of his regularities are mind-bogglingly complex, such as the prototypically "oddly" numbered #37:

> If a numeral expression contains a complex constituent, then the numerical value of the complex constituent itself in isolation receives either simple lexical expression or is expressed by the same function and in the same phonological shape, except for possible automatic phonological alternations, stress shifts, or overt expressions of coordination. (Greenberg 1978: 279–280)

The fact that this rule allows two separate classes of outcome, and three separate classes of exception, suggests that this is not much of a generalization. But other generalizations are exceptionless and nontrivial: for instance, that every language has a finite set of number words, and every language has a highest number beyond which there is no agreed-upon successor. The largest number known to many English speakers is *trillion*, and the largest number in the Oxford English Dictionary, other than the jocular *googol* and *googolplex*, is *decillion*. Beyond that, one can always keep counting, but there will always come a point where one will once again reach the end of the system and need to make up yet another highest number.

Although it is not one of Greenberg's generalizations, one might ask: are there, as Funes the memorious devised, languages in which the word for *five hundred* is *nine*? Or, to put the matter more generally, are there languages that use the same word for two or more numbers? This is very odd to think about—unlike homonyms in other aspects of language, numerical homonyms raise a very strong prospect of ambiguity. Are these homonyms, synonyms, both, or neither? Conversely, is the word for *nine*, *five hundred*? This creates a fundamental semantic ambiguity that leads to surreal statements like *There are five hundred innings in a baseball game.* And yet, surprisingly enough, there are languages that use the same word for two or more numbers. The resolution to the ambiguity is that such languages do not use those words in isolation, but in combination with gestures, or pointing to parts of the body, that make the intended meaning clear: as in "one, another one, another one, another one," raising or pointing to a different finger each time, as Greenberg (1978: 257) postulates for Kaliana (Sapé), spoken in southern Venezuela. Seiler (1995: 144) notes that several languages have the same word for 1 and 6, 2 and 7, etc., because the names of the numerals are the names of specific fingers (as if English used *pinkie* for both 5 and 10). Such languages are rare but highly informative. They highlight the constraint that leads to the generalization—the massive potential for ambiguity that emerges from numerical homonymy—by demonstrating how exceptions deal with that difficulty.

In my earlier work on numerical notations, I investigated the constraints on numerical notation systems in a way partially parallel to Greenberg (Chrisomalis 2010). Numerical notation systems are structured, graphic, principally nonphonetic representation systems for numbers. We're most familiar with the Western numerals 0–9, but also with systems like the Roman numerals, still used for prestige and secondary functions. Numerical notations are not essentially linguistic, although they relate to language—they are a very different *modality* of representing numbers graphically rather than orally. Modalities combine the idea of medium (visual, auditory, tactile, etc.) with the mode by which information is conveyed. So for instance *6* and *six* are both visual, but have different modalities. I return to the topic of modalities in greater detail in chapter 6. While numerical notations frequently co-occur with writing systems, writing is neither a necessary nor a sufficient condition for the existence and widespread use of numerical notation. While tallying, the use of marks, knots, or other signs in one-to-one correspondence with some quantity being counted, may be a precursor to numerical notation structured by a numerical base and its powers, they are not the same. The use of notches or cut marks on bone with a numerical meaning, as is now abundantly attested in the Upper Paleolithic archaeological record and potentially earlier (Marshack 1972; Elkins 1996; d'Errico 1998; Hoffecker 2007), is fascinating and worthy of attention, but is not numerical notation per se, and falls outside this analysis.

More than 100 numerical notation systems are attested to have been used. They range from the very old, such as proto-cuneiform, developed in Mesopotamia ca. 3500 BCE (Nissen, Damerow, and Englund 1993), to the ultra-new—the Inupiaq numerals developed in 1995 by schoolchildren in Kaktovik, Alaska, and used more widely thereafter (Bartley 2002). Some, like the Western numerals, are used by billions; others, such as the Cherokee numerals developed by Sequoyah in the early nineteenth century, which I will discuss in chapter 5, were rejected almost at the time of their invention (Holmes and Smith 1977; Walker and Sarbaugh 1993). Yet all these attested systems are variants of one of five basic types, as shown in figure 1.1 (Chrisomalis 2010: 13).

	Additive	**Positional**
Cumulative	Roman CCXXVIII = 100 + 100 + 10 + 10 + 5 + 1 + 1 + 1 = 228	Babylonian sexagesimal  𒐗 𒐏𒐜 = 3 (x60) + (10 + 10 + 10 + 10 + 1 + 1 + 1 + 1 + 1 + 1 + 1 + 1) (x1) = 228
Ciphered	Greek alphabetic Σ Κ Η = 200 + 20 + 8	Western 228 = 2 (x100) + 2 (x10) + 8 (x1)
Multiplicative	Chinese classical 二百二十八 2 x 100 + 2 x 10 + 8	

Figure 1.1
Typology of numerical notations

Cumulative-additive systems, like the Roman numerals, use a set of signs that can be repeated several times to indicate their addition. Then the signs are strung together, almost always from the highest to the lowest power of the base, and the value of the whole phrase is added up. So, e.g., MMXVIII is 1,000 + 1,000 + 10 + 5 + 1 + 1 + 1 = 2,018. Cumulative-additive notation looks a bit like tallying, but because it uses a base, and because it is not an ongoing count of items, it is really a numerical notation like any other.

Cumulative-positional systems use the same principle of cumulation, by which individual signs are repeated to indicate their addition, but then instead of just adding up sequences of different-valued signs, each group of signs is multiplied by a power value based on its position or place in the entire phrase. This is the concept of place value and originated with the cumulative-positional Babylonian numerals, though today ciphered-positional numerals are more common.

Ciphered-additive systems are ones that have distinct signs for each multiple of each power of the base of the system, which are then ordered in descending order and their values added. So, for instance, the Greek alphabetic numerals use 27 signs for 1–9, 10–90, and 100–900, to write any number up to 1,000. Unlike the Western numerals, this is not a place value system and has no zero, and there is no visual similarity between the signs for, e.g., 4, 40, and 400.

Ciphered-positional systems are historically quite rare, but because the Indian, Arabic, and Western systems in common use today have this structure, it's easy to take them for granted. Structurally, they have separate signs for each number from 1 up to the base, and then these signs take on an implicit power value based on their position or place in the entire phrase. Most ciphered-positional systems use a zero to indicate a blank or empty place, for clarity.

Multiplicative-additive systems, like the classical Chinese numerals, normally use two signs for each power of the base, which are multiplied together: a multiplier, representing a number from 1 up to the base, and a power sign, representing the specific power.

These five types can sometimes be combined in interesting ways, and there are other differences (some systems use base 20 or 60, as in the Babylonian system above), but these five basic principles recur again and again cross-culturally, and no others. You can imagine other systems quite easily. For instance, you could have a system where the size of a sign indicates its value—so that a big 2 might represent 20 and a really big 2 might represent 200 or 2,000. Or you could have a system where every number is written as the product of prime numbers, because each natural number has a unique prime factorization—e.g., 20 would be 2 2 5. Or a system that has a different base for each power—so that 225 might mean two twenties, two sevens, and five ones (i.e., 59). All of these systems are coherent—they would let you write any number starting with 1 in an unambiguous way. But such aberrant systems were never used by anyone as far as can be ascertained.

There are powerful cognitive constraints limiting the possibilities for actual, workable numerical notation systems (Chrisomalis 2010: 360–400). Among attested numerical notation systems, there are 18 strong regularities, a few of which are universal but most of which, while generally valid, have one or more exceptions. Some of these follow Greenberg's regularities for lexical numerals quite closely: there is no numerical notation system that just skips a number, for instance. Such a thing is not impossible to imagine. Karen Blixen (better known by her pen name, Isak Dinesen) imagines just such a thing in her memoir *Out of Africa*, with her story of a translator, embarrassed because the Swahili word for *nine* had a "dubious ring" in Swedish, who tells her that Swahili has no word for nine nor any compound of nine, although it had all the other numerals (Blixen 1937: 263–264). However, her wonderment and bemusement at this possibility serve to demonstrate just how unlikely it is that such a circumstance could actually arise:

> At the time when I was new in Africa, a shy young Swedish dairy-man was to teach me the numbers in Swaheli. As the Swaheli word for nine, to Swedish ears, has a dubious ring, he did not like to tell it to me, and when he had counted: "seven, eight," he stopped, looked away, and said: "They have not got nine in Swaheli."
>
> "You mean," I said, "that they can only count as far as eight?"
>
> "Oh, no," he said quickly. "They have got ten, eleven, twelve, and so on. But they have not got nine."
>
> "Does that work?" I asked, wondering. "What do they do when they come to nineteen?"
>
> "They have not got nineteen either," he said, blushing, but very firm, "nor ninety, nor nine hundred,"—for these words in Swaheli are constructed out of the number nine,—"But apart from that they have got all our numbers."
>
> The idea of this system for a long time gave me much to think of, and for some reason a great pleasure. Here, I thought, was a people who have got originality of mind, and courage to break with the pedantry of the numeral series.
>
> One, two and three are the only three sequential prime numbers, I thought, so may eight and ten be the only sequential even numbers. People might try to prove the existence of the number of nine by arguing that it should be possible to multiply the number of three with itself. But why should it be so? If the number of two has got no square root, the number of three may just as well be without a square number. If you work out the sum of digits of a number until you reduce it to a single figure, it makes no difference to the results if you have got the number of nine, or any multiple of nine, in it from the beginning, so that here nine may really be said to be non-existent, and that, I thought, spoke for the Swaheli system.
>
> It happened that I had at that time a houseboy, Zacharia, who had lost the fourth finger of his left hand. Perhaps, I thought, that is a common thing with Natives, and is done to facilitate their arithmetic to them, when they are counting upon their fingers.
>
> When I began to develop my ideas to other people, I was stopped, and enlightened. Yet I have still got the feeling that there exists a Native system of numeral characters without the number of nine in it, which to them works well and by which you can find out many things.
>
> I have, in this connection, remembered an old Danish clergyman who declared to me that he did not believe that God had created the Eighteenth Century.

In other ways, however, the set of regularities in numerical notation diverges from Greenberg's considerably. For instance, while all lexical numeral systems have a limit beyond which newly invented words are needed to extend the system, many numerical notation systems are truly infinite: they consist of a small set of symbols and a rule set allowing the construction of a numeral representation of indefinite length, without ever having to develop a new symbol. In decimal numerals with place value like Western numerals, you can simply add another zero. It is certainly imaginable that English could have a lexical numeral system that simply replaces the digits with the words *zero* through *nine*, and thus would become infinite; for instance, if we simply said *two zero six three* instead of *two thousand and sixty-three*. And, of course, English speakers do sometimes use such readings: for instance, when reading phone numbers. However, these digital readings are secondary and derive *from* graphic notation. In English, they emerged only in the late nineteenth century, with the rise of telephone numbers and other technologically influenced numbers. No language has as its basic, ordinary number system anything like this—even though it could, very readily.

Similarly, while many lexical numeral systems are not ordered strictly by descending order of powers, numerical notations virtually always are strictly ordered. Strict ordering of linguistic numeral phrases by descending powers is common, and Miller and Zhu (1991) show that children acquire numeral words in languages with strict power ordering (such as Mandarin) somewhat more quickly than in languages with irregularities (such as English). Yet there are many exceptions to this linguistic generalization—e.g., German *vierundsiebzig* (4 + 70) for 74 instead of **siebzigundvier*, or in English *sixteen* (6 + 10) instead of **ten-six*. But in numerical notation systems, strict ordering by powers is virtually always followed.[4] In English, at least, the divergence between the two representations of numbers is best explained by the fact that lexical numerals frequently have two different morphemes representing the same number, such as English *-teen* and *-ty* which both mean *ten*. The difference is not in the number they represent, but the arithmetical operation they signal. *Thirteen* and *thirty* are both morphologically "3 10," but *-teen* indicates that a number from 11 to 19 is to be formed through addition (3 + 10), whereas *-ty* indicates that a decade from 20 to 90 is to be formed through multiplication (3 × 10). In numerical notation, where such morphological variability does not exist, strict ordering is necessary to avoid ambiguity. Numerical notations consist of small sign inventories of distinct graphemes, and to have multiple signs for 10 would defeat one of their key functions, to communicate numbers rapidly with discrete and easily differentiable signs. Thus, because of their graphic modality, numerical notation systems need power ordering to a greater degree than lexical numeration does.

The graphic modality of numerical notation also helps to explain a major structural difference from lexical numerals, namely the much greater role of sign repetition. Repetition (e.g., reduplication) is quite rare in lexical numeral systems; so, for instance, there is no language where *three hundred* is expressed as *hundred hundred hundred.* In contrast, repetition of signs to indicate addition is extremely common in numerical notation—it is the fundamental principle of cumulative-additive systems such as the Roman numerals and the Egyptian hieroglyphs and also cumulative-positional systems such as the Inka *khipu* numerals and the Babylonian sexagesimal numerals. Virtually all lexical numeral systems use multiplication using powers of one or more numerical bases ("three hundred"), and while this is sometimes used in numerical notation, it is not the predominant representational mode. Here, the constraint is the visual nature of graphic representations in contrast to the vocal-auditory channel used to speak and listen. Not only is "ten ten ten ten ten ten one one one one" much longer than 64; because of the impermanent nature of vocal-auditory signals, such numerals would be difficult for a listener to comprehend and retain, or to distinguish from, say, 54 or 65. With numerical notation, in contrast, because graphic notations are relatively permanent and can be examined in their totality by the reader, a notation like XXXXXXIIII is not too difficult to perceive. We will examine the relationship between conciseness, repetition of signs, and cognition in the next chapter.

For generativist linguists, universals of verbal numerals (and universals of language more broadly) help to demonstrate that specific aspects of language are "hardwired" in the human brain. If they are hardwired, they must be evolutionarily as old as anatomically modern humans, many tens of thousands if not some hundreds of thousands of years old (Wiese 2003).[5] However, regularities in numerical notation also exist, some in parallel with and some divergent from those that exist for lexical numerals. Because there was no numerical notation before 6,000 years ago, when the earliest systems developed alongside early writing, this is far too recent to be an evolved capacity, and the functional association with the origins of writing is not coincidental (Schmandt-Besserat 1992). There are universals, near-universals, and statistical generalizations about numerical notation, just as there are for lexical numerals. In other words, we have two sets of different universals for the domain of number, one of which cannot possibly be biologically innate. There are certainly innate evolved capacities that relate to numerical notation, but these are not sufficient, by themselves, to explain the regularities in question. And if there is no possibility that the regularities of numerical notation are fully innate, there is no reason to assume that the regularities of lexical numerals are fully innate. If there is a universal grammar for lexical numerals, therefore, it is quite plausibly universal in part because of the kinds of nonbiological constraints that produce regularities in numerical notation, rather than being narrowly "hardwired" or located solely in the brain.

Because number is a domain where there are two or more different representations in many different societies, we can investigate why some regularities might be true of number words but not of number symbols, or vice versa. For numerical systems, one of the most crucial constraints is to avoid ambiguity. In most situations, where there are few contextual cues to distinguish the meaning of one number from another in an utterance or text, it is vital to avoid ambiguity. Combining a verbal or graphic representation with gesture, as described above, is one solution. Consistently using a limited set of graphic signs, ordered from highest to lowest power, is another.

Comparing these two modalities, we see some constraints that result in true universals across both modalities, others that result in universals in only one or the other but not both, and many others in which we see patterns and strong generalizations but also exceptions in both. The question of how variable any particular domain is, then, is interrelated with the question of the number and strength of constraints operant in that domain. How do constraint theories help resolve the problem of exceptions within general patterns?

The 99% problem

The linguists Nicholas Evans and Stephen Levinson have issued a major set of statements against universalism in linguistics that challenges the predominant view that linguistic universals are numerous and important (Evans and Levinson 2009; Levinson and Evans 2010). They argue that despite over a century of serious work on the question, the number of true, exceptionless universals of language is vanishingly small. And yet, far from retreating into particularism, Evans and Levinson note that a wide variety of features are common to virtually all, but not quite all, known languages. Everett's example of numeralless Pirahã is one of them, but certainly not the only one. For instance, Margetts (2007) shows that Saliba, a language of northern Colombia, violates the near-universal that the verb "give" takes three arguments—in other words, we expect "give" to require both a giver and a recipient as well as an object (X gives Y to Z) to be grammatical. The study of the rarely attested is a small but important area in contemporary linguistics (Pericliev 2004; Wohlgemuth and Cysouw 2010).

This is what we might call "the 99% problem." If something is true of 99% of languages, or 99% of numerical notation systems, or 99% of screwdrivers, but there are one or two exceptions, what do we do with this fact? For a universalist, the answer is easy: ignore the exception. After all, we all ignore exceptions and outliers in our data all the time. For a particularist, the answer is also easy: use the exception as a demonstration of variability in the face of the apparently universal. But this hardly seems satisfactory as an accounting of the degree of patterning in human societies. For a comparativist, however, such a situation invites an opportunity—and a challenge.

For Evans and Levinson, near-universals (strong but not quite exceptionless regularities) are evidence of "stable engineering solutions satisfying multiple design constraints, reflecting both cultural-historical factors and the constraints of human cognition" (Evans and Levinson 2009: 1). For any phenomenon you might want to look at, there are multiple constraints, some of which are likely to be in conflict with one another, and multiple solutions, which weight some of those constraints more heavily than others. For each phenomenon, we must initially remain agnostic about whether these constraints are likely to be stringent or lax, whether they will be numerous or few, and what particular form they will take. We must also remain aware that any situation with multiple constraints is likely to produce exceptions, and must take those exceptions seriously in our analysis.

Within this sort of constraint theory, in the literal and original sense of the phrase, the exception proves the rule. This doesn't mean "prove" in the way we usually mean it in English today—to demonstrate it to be true—but rather, "prove" as in "proving ground"—the place where something is tested. The Latin version of the saying is *exceptio probat regulam*—"the exception *probes* the rule" might be a better translation. It neither demonstrates the rule nor denies it—rather, it clarifies its scope and confirms its general validity otherwise. So, for instance, if you park your car beside a sign that says "No parking 8:30 am–5:30 pm," the time prohibition equally confirms that at other times, parking is permitted. And if you were to see that sign and think to yourself, "Why those times?" you would probably be able to reason that those are business hours. Similarly, for cultural phenomena, analyzing exceptions to general patterns is a powerful analytical tool for understanding why the pattern emerges in the first place. The exception probes the rule, and also helps rationalize its existence, by explaining cases when it doesn't apply.

Exceptionless universals, where they exist, are important but do not provide any clue as to why they might be true. In Donald Brown's important study of human universals, he notes quite rightly that universals are not necessarily hardwired or inevitable—for instance, that there may be "new universals" such as tobacco and metal tools, as well as "former universals" such as breastfeeding infants (Brown 1991: 50). But we only know that these categories are interesting because we are aware of attested societies (past or present, respectively) where the universal does not hold. Thus, near-universals—widespread generalizations with a few exceptions—are extraordinarily important (Dryer 1997, 2003), because they allow anthropologists and linguists to probe for explanations without having to go to other disciplines (such as psychology and biology) to explain the broader patterns being observed. I thus disagree with Brown's (1991: 45) claim that "the distinction between a near-universal and a universal (or absolute universal) ... is not significant, or imposes an artificial break in an unbroken natural continuum. A near-universal is universal enough." This reasoning implies that exceptions are uninteresting or meant to be ignored.

When evaluating rarities in numerical notations, a further consideration is that some kinds of numerical notation systems are easy to imagine, and in fact have been invented, but their structures lead them to be quickly rejected. Many of these systems in antiquity simply will not have survived: their rarity makes their archaeological or archival preservation unlikely. When we look in the modern era, we find many such systems, simply because preservation is so much better. Take, for instance, the *Codex Seraphinianus*, Luigi Serafini's fictional surrealist encyclopedia (Serafini 1983). Its script, though presenting the appearance of decipherability, is meaningless, and it is filled with arcane diagrams of impossibilities. Yet its written numerals can be securely deciphered, for two reasons. First, the context is a set of page numbers, which really do appear to start with 1, at the bottom of every page, and the work has a table of contents that uses these numbers. Second, the system has structure—it repeats on a more or less consistent basis and thus allows a user to fairly reliably predict what will follow on the next page (Derzhanski 2004). Yet the *Seraphinianus* numerals violate several of the regularities that apply to numerical systems that are used for any amount of time in actual social contexts (for instance, the system has a base of 21).

It appears, then, that the constraints on human imagination are far less than the constraints on human behavior. If we had perfect knowledge of the past, we might find dozens, possibly even hundreds of structurally anomalous numerical notations that were rapidly invented and rapidly discarded. The constraints that move us from the imaginable to the attested are primarily cognitive and functional. Only the five general types of numerical notation are, as Evans and Levinson call them, "stable engineering solutions." Systems that diverge from these significantly either rapidly converge to one of those five, or else disappear entirely. There is still plenty of variability, because humans are relatively flexible. As the social anthropologist Christopher Hallpike puts it, it is not about survival of the fittest, but rather "survival of the mediocre" (Hallpike 1986: 81–145).[6] There is no one optimal solution but a limited set of workable solutions, and only the considerably unworkable are discarded, unused.

Constraint, history, and cognition

But, having shown that there are constraints that operate on the world's numerical notations that affect what structures are developed and thrive, we face another, probably even more critical issue. What do numerical notations *do* for their users? Or, put in the terms of this chapter, what are the affordances of any particular numerical notation, and how does it constrain the way its users employ it and engage in numerical practices? This flips the question of constraints on its head—asking not what constrains numerical notations, but what they constrain.

This issue parallels the debate in studies of literacy in the 1960s to the 1990s with respect to the consequences of literacy for individuals (Goody 1977; Scribner and Cole 1981; Olson 1994). The social anthropologist Jack

Goody, in the 1960s through the 1980s, put forward a set of proposals, not always fully congruent with one another, focused primarily on the question: In what ways or in what contexts does becoming literate make a cognitive difference for an individual? For Goody, part of what made the difference was the particular notations being used—for instance, he believed, at least initially, that learning an alphabet had different cognitive consequences for its users than learning a nonalphabetic script (Goody and Watt 1963). So, for instance, there is evidence that alphabetic literacy makes users more aware of some phonetic aspects of their languages than literacy in nonalphabetic scripts (Olson 1994).

However, literacy is not simply about the code you learn, but how you learn it and how you use it. Sylvia Scribner and Michael Cole, in a superlative ethnographic and psychological study of over a thousand speakers of Vai, a Niger-Congo language of Liberia, showed that the situation is far more complex (Scribner and Cole 1981). Vai literacy is unusual because there were, at the time of their study and still to some degree today, three scripts regularly taught and used: the Roman alphabet, the Arabic abjad (primarily consonantal with some vowel marking), and the locally developed Vai syllabary, where each character represents a syllable rather than a single sound.[7] Scribner and Cole showed, in a set of ingenious experimental tasks, that what mattered most was not the particular features of the script itself, but how it was learned and how it was used. In other words, they provide a practice-based account of literacy and its cognitive consequences. Particular forms of schooling were very important, and while those forms of schooling involved literacy, simply being literate was not the critical element. As it turns out, literacy may be necessary for certain sorts of cognitive changes in humans but is certainly not sufficient. And the form of script that is used doesn't seem to matter in the way that we might think—the more radical, deterministic reading of Goody's work can't be true. This is particularly pertinent for discussions of East Asian scripts, which are often regarded as inferior because they encode morpheme-level or word-level information to a much higher degree than they do phoneme-level information—although in reality they are far more phonographic than often believed by nonspecialists (DeFrancis 1984).

Similarly, because numeracy is a cognitive activity, and numerical notation is a cognitive technology, we might imagine that people who use Roman numerals, for instance, would think about number differently than people who use Western numerals. But because there are no "native speakers" of Roman numerals today, it's not possible to do the kind of natural experiment that Scribner and Cole engaged in. Pretty much all literate people today are familiar with, and use, ciphered-positional systems like the Western numerals. In East Asia, where multiplicative-additive numerals are learned very early and used with some regularity, mathematical activity is either done using ciphered-positional numerals or with the aid of an abacus (Chinese *suan pan*; Japanese *soroban*). There is some really interesting cognitive-psychological work on East Asian abacus experts, showing that they do mental arithmetic with a "mental abacus"—a sort of representational structure that, even in the absence of the object itself, gives them a scaffold for performing arithmetic (Stigler 1984; Hatano and Osawa 1983). This suggests that, if we were able to go back in time and test users of other numeral systems, we might find similar and interesting cognitive effects—at least, in the context of numerals used for arithmetic. But note that the work on abacus cognition has been done with experts—people with years, even decades of practice with the technology. So it is wrong to think that the tool alone causes an effect—the tool is integrated in an entire technological system that includes tools, minds, problems, and contexts.

Similarly, just as a toddler and a topologist may both use the same set of numerals 0–9, but with enormously different cognitive outcomes, we should not presume that any particular notation automatically or universally has any particular cognitive effect. Once again, we return to the definition of constraints as processes that limit but do not specify a particular outcome. We can then analyze how variability in their features is selected for or against in particular contexts of use, and understand how, within the general constraints offered by universal cognitive processes and environmental preferences, specific numerical notations may flourish (or not) under particular social and historical constraints.

Let us return to the Roman numerals for a moment, and consider the widespread assumption that they were replaced because they are characteristic of an earlier stage of human cognition—as Dehaene (1997: 66) calls them, a "living fossil"—or at worst, as Murray (1978) asserts, that they actively inhibit mathematical cognition. This claim is seen frequently in the cognitive literature on numeracy (e.g., Zhang and Norman 1995) under the guise of the assumption that it is reasonable to compare numerical notations on the basis of their efficiency for written computations. Even if we grant that Roman numerals could not have been used as the basis for higher mathematics—a claim on which Schlimm and Neth (2008) cast some doubt—they thrived in Western Europe for over two thousand years, including five hundred years (roughly 1000–1500 CE) when they coexisted with the Western numerals without replacement. To know why the Roman numerals were replaced, we cannot simply compare their structure to that of their successor, the Western numerals, but we must also ask for what functions they were used. For instance, the Roman numerals were never used for written arithmetic, as the Romans had the perfectly serviceable pebble-board abacus in its place.[8] Only when written arithmetic, associated with widespread literacy and the printed book, became feasible and common was the Roman system vulnerable to being replaced in Western Europe; it was simply too well entrenched and the modes and technologies of its use too well known. For any communication system to be useful (whether an alphabet, or a music notation system, or a system of traffic signs), it needs to be shared in common by a body of users, and one reason a system may be rejected is that it is perceived to be awkward for some function, which might be arithmetical but might be something else entirely. In the case of the Western numerals on their first arrival in Western Europe, xenophobia may have played a role, but zero was also considered particularly enigmatic and difficult to work with.[9] Another factor is that abandoning a preexisting system would mean you could no longer communicate well with those who do not know the new system. In such cases, either factor would be a constraint, but we cannot predict which would be most important in any particular case. In chapters 3 and 4, I will return to this argument at greater length.

I have serious doubts that it is ever possible to predict the cognitive effects of a numerical notation directly from its structure. The Assyriologist Peter Damerow's (1996) careful work on the developmental psychology underpinning the archaic Mesopotamian numeration (from the late fourth to the mid third millennia BCE) is the most consistent and thorough attempt to argue that the stages of the evolution of notation parallel cognitive stages. He argues that the presence of multiple (perhaps as many as 30) different numerical notations in the Uruk period (ca. 3000 BCE) should be seen as evidence of absence of abstract number concepts, because there was no way to refer to an abstract number without simultaneously designating some concrete object being enumerated (grain, cattle, people, etc.). However, the lexical evidence from Sumerian suggests that there was, even at this early date, a single, perfectly ordinary set of number words (Powell 1972). Taken literally, this would imply that nonliterate Sumerians with their single abstract number system were more abstract in their numerical cognition than literate ones—a position Damerow surely does not wish to take. A far more reasonable explanation is that, just as we count time in base-60 and use hexadecimal (base-16) notation in computing, the technical needs of administering the Uruk city-state made it advantageous for Sumerians to use different notations when employing different systems of weights and measures.

However, to say that one cannot simply read cognitive effects from written number notations is not to say that there are no such effects, but rather to demand further empirical investigation, including social context. Jack Goody (1977: 85–89) rightly emphasizes the importance of numbered lists as a cognitive tool for organizing information that better permits the visual inspection of knowledge. While arithmetic is cross-culturally quite a rare function of written numerals, and thus not a good candidate for identifying cognitive effects, list making is a widespread phenomenon in which written numerals are employed. Maybe number systems that more readily allow the reader to evaluate the size of numbers are better suited for such a function. Now we have a hypothesis that we might be able to test. Because every three-digit Western numeral is larger than every two-digit numeral, one can use the length of numeral phrases as a rapid proxy for numerical size. For Roman numerals, it is not immediately apparent that CXXI is larger than CXVIII. While this does not explain the replacement of Roman numerals, it does suggest fruitful avenues for research. Because the cognitive literature to date has focused almost exclusively on Western numerals, there has been no opportunity to consider these potentially important effects.

In a global network of societies in which there is a strong interest, particularly among elites, in science, commerce, technology, and education, social and communicative constraints make the adoption of a standard notation to facilitate social networks very likely, though not inevitable, and even systems of great longevity can be replaced very rapidly. This is not only true for the Roman numerals but for many other systems that were rendered obsolete at the advent of the modern world system, between 1450 and 1650 (Chrisomalis 2010: 423). This was a period which saw a wide range of newly literate users of written numbers (e.g., the mercantile professions) choosing the Western numerals in part because of their utility for pen-and-paper computation, and the role of written arithmetic texts in transmitting this information. But it would be a mistake to imagine that the Western numerals were developed *for use in* arithmetic, just because they afforded certain new arithmetical practices. They are best seen as an instance of *exaptation* (Gould and Vrba 1982)—in evolutionary terms, something that developed for one purpose but came subsequently to serve another. An abacus is a remarkable computational tool, but it leaves no trace of immediate steps in computation, which is hardly advantageous to commercial enterprises on a national or global scale. The fact that Roman numerals are unsuited for written computation was irrelevant until such time as written arithmetic became a highly desirable function, in early modern capitalist economies. Once that had occurred, however, the Roman numerals declined in use quite rapidly. In chapter 4, I return to the relationship between frequency of use and perceived usefulness of notations.

The historical pattern of invention and decline of numerical notations suggests that in some key ways, the past five hundred years (or, should I say, following Borges's Funes, nine years? or, then, following Dinesen's Swahili translator ... say nothing at all?) may be unlike the preceding five thousand. Most notably, the massive integration of societies has made it very difficult for novel numerical systems to gain a foothold due to these newly important social constraints, which exist alongside cognitive and graphic/notational ones. Yet, even though numerical notations are disappearing more rapidly than in the past, the systems that remain, and recently invented and used systems, conform generally to the structural principles observed in the premodern world. To abandon the study of the cognitive and structural constraints governing written numerals, and to view the choices underlying the adoption, transformation, and decline of numerical notations (and scripts) as motivated solely by social constraints, would miss the point. Numerals are used by people, in real social contexts, for a variety of reasons. Once we are aware of this, we can integrate the two perspectives outlined at the beginning of this essay: how cognitive factors constrain numerical systems, and how numerical systems constrain cognitive outcomes. The next and vital step is to extend this work cross-culturally and cross-linguistically to generate testable general theories that consider cognition and representational systems as a central aspect of the human experience of mathematical ideas.

2 / II Conspicuous computation

Written numbers are a part of almost every contemporary human environment. Take a moment to look around the room where you're reading this page. They are everywhere. Perhaps stamped underneath your comfortable chair, a serial number. Around your wrist, a number telling you how many steps you have taken today. Inside a wedding ring, a carat mark, 14. On TV, a batting average, a box score, the top of the seventh with your team one run down. For the most part, we don't think deeply about these numbers, even though it is certain that people in the industrialized world encounter hundreds of them each day. They linger below awareness.

But now I want you to think of each of these numbers, not as some randomly distributed feature of your landscape, but as an intentionally created product of a literate and numerate mind—perhaps not with you, *specifically*, in mind, but with some audience in mind, and some purpose underlying a decision to write that number, in that way, at that time. In other words, to think about written numbers as texts. How were these numbers chosen? Why does your credit card have sixteen rather than some other number of digits? Why do we separate every third digit in a number greater than 1,000 with a comma—and why do we not do so with numbers to the right of the decimal point?[1] Why are there Roman numerals numbering the pages of the front matter of this book? And why, after all, don't we use the Roman numerals much anymore?

The Roman numerals were nearly ubiquitous throughout Western Europe from the third century BCE to the fifteenth century CE. They fell out of use in the early modern period: retained for prestige or archaic functions, or where a second set of numerals is useful, but abandoned for ordinary purposes. If I were to ask you why, as I have asked hundreds of people over the years, you might say something like, "Have you ever tried to *multiply* with those things?" An understandable response. Many of us may have a memory of being taught Roman numerals in school and being given artificial puzzles like DCXXXVII – XCIII: a minor form of torture, perhaps. Roman numerals, to the mind trained in modern Western arithmetic, may seem obscure, awkward, and simply too *long* to be of use. But not only don't we do arithmetic with Roman numerals, neither did the Romans or the medieval inheritors of their notation: they used an abacus or counting-board for calculation, and used written numerals only for representation. A better approach to this question would ask about the choices and motivations of writers and their audiences in relation to the texts themselves, within specific social contexts. Whether we write 637 or DCXXXVII or εξακόσια τριάντα επτά is not just a matter of what language we speak, but what audiences we imagine ourselves to be addressing and what impact we want to have on them.

This chapter builds on the framework of cognitive constraints and patterns outlined in the previous chapter, framing the decisions that writers and readers make as rational, understandable responses to communicative challenges. I will introduce a set of texts from around the world that seem almost flagrantly to deny the general norm that shorter is better when it comes to numbers. But to get there, I need to outline a framework for understanding how numerical texts create meaning. From there, I conclude with some insights into the problem of the Roman numerals that bear on broader questions in social and linguistic analysis.

Dynamic philology

Relatively few people today are deeply familiar with the discipline of *philology*, the study of the historical, linguistic, and literary aspects of texts. Philology encompasses interpretive and semiotic aspects of both the languages and literatures of the premodern world. In its heyday in the nineteenth century, it was one of the central disciplines of what would become the modern European university. Its influence persists across disciplines such as linguistics, anthropology, comparative literature, and even cognitive science, even though today there are only a handful of actual departments of philology remaining in academia. Pollock (2009) argues persuasively that while philology is largely the subject of scorn in recent scholarship, it has an important role to play in helping make sense of texts in contemporary thought. Philology is a strange *-ology*; for every other "-ology" branch of study, the second morpheme, *logos* "word," is used metonymically to stand for the study of some thing—so, for instance, sociology, words about society, hence the study of society. In *philology*, though, *logos* is literal—it means "word," and philology is thus the love of words. As a linguistic anthropologist, and thus one of the intellectual inheritors of philology, I find it a compelling framework (Hymes 1963; Bauman 2008). Treating written numbers not just as a set of signs but in terms of their linguistic history and social context allows us to ask and answer new questions and solve some otherwise strange-seeming puzzles.

The study of numerals has long been a subject of philological interest. By the nineteenth century, the stability and frequency of numerals in Indo-European languages helped them to serve as a model domain for understanding the family as a whole. To this day, one of the principal divisions in Indo-European historical linguistics is between the *satem* and *centum* languages, two subgroupings of Indo-European whose distinction was drawn after the form of words for "hundred" in Avestan (an ancient language of Iran) and Latin (Renfrew 1990: 106–108), respectively. This division helped us to understand the history and processes of change in a language family encompassing hundreds of modern languages. The first book-length philological study of numerals was August Friedrich Pott's *Die quinare und vigesimale Zählmethode* of 1847, in which he compared quinary (base-5) and vigesimal (base-20) number word systems worldwide. Before Pott, while there was certainly some awareness that the number words of the world's languages were variable, no one had really tried to explore that variability systematically. Pott showed that while base 10 (decimal) notation was common, these other structures, previously understudied and certainly not compared to one another, were geographically widespread. A decade or so later, Antoine Pihan's *Exposé des signes de numération* (1860) was the first systematic comparative study of systems of graphic number symbols, including both the ancient systems of the Near East as well as the contemporary systems of Europe, South Asia, and East Asia. These two subjects of study, the comparative study of lexical numerals (number words) and the comparative study of numerical notations (number symbols), remain at the heart of the philological enterprise. Numerals are important both as linguistic phenomena and as graphic ones, and, because they represent the universal human domain of number, are amenable to cross-cultural analysis. They bear the impressions of both the cognitive properties of the mind and the local linguistic and social practices of specific times and places.

Philology no longer occupies the esteemed role that it did in the Victorian era, and it is often thought of as a dead discipline. But rather than writing its eulogy, I think it is better to talk about its *dissemination*: where once there was a single discipline, we now have a set of problems and challenges spread across linguistics, comparative literature, linguistic anthropology, semiotics, and the various historical and classical disciplines, each of which brings to bear its own methods and theories to philological questions. The seeds planted in the dissemination of philology have borne fruit in my own discipline, linguistic anthropology, that we can use to address questions in the study of ancient and modern number systems. To do so, I want to draw in one of the rather different fields in which these seeds found fertile ground: the quantitative linguistic approaches pioneered by George Kingsley Zipf.

Zipf (1902–1950) is not today thought of as a philologist. He is best known for two contributions to the language sciences, and more broadly to information science. The first of these is the eponymous *Zipf's law*. Zipf demonstrated that across a wide range of domains of activity, there is a power-law distribution of frequencies of ranked phenomena. For instance, across many languages, the most common word (in English, *the*) is twice as frequent as the second most common, ten times as frequent as the tenth most common, and so on. But also, strangely, across many countries, the most populous city is roughly ten times as populous as the tenth most populous city, and one hundred times as populous as the hundredth on the list. That these power-law distributions of ranked phenomena are widespread across both human and nonhuman phenomena is now incontestable, although their explanation is hotly debated still (Newman 2005).

Zipf's second major contribution to scholarship was what he called the *principle of least effort*, a term which has antecedents in the French philosopher Guillaume Ferrero's 1894 paper "L'inertie mentale et la loi du moindre effort" (Ferrero 1894) but which took mathematical form in Zipf's empirical work (Zipf 1949). Zipf set forward a general behavioral principle that, all other things being equal, humans will prefer solutions that minimize effort (even at the expense of other considerations). This is more than just the truism that humans are lazy. For instance, applied to language, Zipf's prediction, borne out across many languages, is that there is a strong inverse correlation between word frequency and word length, so that most of the frequent words in any language will have fewer syllables than the rarer ones. Almost all the pronouns, prepositions, conjunctions, and other so-called "function words" have one syllable in English, and are also among the most common words in any conversation or text. Zipf's law of least effort applies broadly across many languages that have been investigated in this way.

It is probably a fair assessment to say that, on an interdisciplinary scale, Zipf has been the most influential philologist of the twentieth century, and now into the twenty-first. This influence, however, has been almost exclusively outside of the humanistic descendants of nineteenth-century philology that I mentioned earlier. Instead, Zipf's publications are most widely cited and used today in computer science, cognitive science, engineering, and demography. Even in linguistics, his influence is largely in computational and corpus approaches where "big data" rules the day (Montemurro and Zanette 2011). But his training was grounded in Harvard's distinguished program in comparative philology, where he studied for the doctorate under Joshua Whatmough, who was central in the transition of philology to contemporary linguistics at Harvard, and who pioneered the use of statistical methods in the study of Latin texts. Both Whatmough and Zipf were iconoclastic, but the latter far more so. While at Harvard, Zipf developed, in his dissertation, ideas that would become what he came to call "dynamic philology." The major part of his dissertation, very unusually, was published as a single 95-page article in the august *Harvard Studies in Classical Philology* (Zipf 1929). While it had almost no impact at the time (I can find only four references to it from the decade after its publication), it is now the most-cited paper ever from that journal—as of 2020, 323 of its 382 citations are from the twenty-first century. What a wonderful object lesson (in case you ever need evidence in support of a tenure case) that lack of immediate "impact" does not entail irrelevance! Zipf then developed these ideas in fuller detail in his 1935 monograph *The Psycho-biology of Language*. He does not mince words; the first sentence of the book is "Dynamic Philology has the ultimate goal of bringing the study of language more into line with the exact sciences" (Zipf 1935: 1). At times he seems almost resolutely opposed to any interests we might regard as philological.

I should disclose my biases up front: I believe Zipf's *The Psycho-biology of Language* to be an extraordinarily important piece of linguistic scholarship, one whose central findings about word frequency have been replicated across many languages and which have applicability in natural language processing, cryptography, and corpus linguistics. Yet I do not think it is a mistake that twentieth-century philologists took relatively little note of Zipf's work, in large part because Zipf showed almost no interest in their questions, approaches, or even subject matter. It is striking that his own doctoral advisor, Whatmough, never mentioned or cited Zipf's work in print until 1952, two years after his student's death (Whatmough 1952). The challenge is not, I think, primarily with Zipf's quantitative focus—while Zipf was one of the first to systematically use large corpora for philological purposes, and to use statistical methods in the analysis of texts, I would think that its use is now sufficiently widespread across digital humanities, archaeology, and epigraphy that I need not belabor its utility.

Rather, there are at least three features of Zipf's body of scholarship that militate against his influence in philology. The first is that his dynamic philology was resolutely universalist: Zipf saw speech patterns as grounded in human psychology and biology, and sought to generate lawlike principles governing language that would be applicable to every language and context. So, for instance, his principle of least effort saw brevity as a universal human goal, counterbalanced only with the need to avoid ambiguity. While I would expect that there would be general agreement to the idea that there is a biological and psychological basis to language, Zipf showed practically no interest in anything that might vary cross-linguistically or cross-culturally, or might apply differently in different contexts. As I discussed in the previous chapter, while I am very interested in cross-cultural patterning, the narrow focus on universals is rarely of direct scientific utility—it is, rather, through examining patterns with exceptions using comparative methods that we gain a richer understanding of underlying principles.

Secondly, and relatedly, Zipf's approach was staunchly positivist. Those of us who grew up during the "theory wars" of the 1990s have perhaps become accustomed to the use of "positivist" as epithet. Here I mean it in the narrowest of senses: for Zipf, speech, as behavior, served as the observable evidence on which generalizable conclusions were based, and Zipf was deeply reluctant to go beyond word frequencies themselves in attributing causation to his well-established correlations. While acknowledging that "the qualities of intelligence, value, and experience are especially vital factors in speech behavior," he insisted on using quantitative tools "without any reference to these seemingly variable and highly elusive factors" (Zipf 1935: 11). This was typical of the anticognitive biases of the period in which he wrote, but again is antithetical to my approach, which regards human cognitive constraints and capacities as fundamental to a deeper understanding of the patterns that Zipf (and many others) found.

Finally, and strikingly for a philology calling itself "dynamic," Zipf's data and approach relied almost solely on synchronic data—those that examine language at a single point in time, as opposed to diachronic data that examine change over time. Zipf was interested in languages as the product of dynamic, equilibrium-seeking systems, which justified him treating them as the end products of those processes through inference, without considering actual, observed change. Any one of these perspectives might have been enough to earn the disregard of philologists; all three, taken together, put Zipf well beyond the pale. A pair of scathing reviews of his work in the journals *Language* (Joos 1936) and *American Anthropologist* (Swadesh 1936) put an end to any potential influence Zipf's approach might have had at the time. As someone whose work is grounded in historical, diachronic analyses of change, I similarly find Zipf's dynamic philology extraordinarily ahistorical in places.

So what do Roman numerals have to do with Zipf's theories? And how, more generally, might we regard "dynamic philology" in the twenty-first century? Despite the rather dramatic gulf between the traditional studies of meaning in the humanities and what Zipf proposed, I think we can learn a great deal by integrating the two. But to make this case, I will need to dive back into a broader range of comparative examples from numerical systems, to show exactly where Zipf's principles do *not* apply.

Conspicuous computation

The so-called Narmer mace head is one of a small corpus of Egyptian hieroglyphic and proto-hieroglyphic inscriptions from the Early Dynastic reign of Narmer, roughly the thirty-first century BCE (figure 2.1). Found at Hierakonpolis (ancient Nekhen), it is a decorative stone mace head that was probably a royal gift to the temple there, depicting key activities of a particular year of Narmer's reign in bas-relief (Quibell, Green, and Petrie 1900; Millet 1990). It is also the only Early Dynastic Egyptian text that provides any evidence for the numerals for 100,000 (a stylized tadpole) and 1 million

Figure 2.1
Narmer mace head (Budge 1911: 36)

(the representation of the god Ḥeḥ, the personification of infinity), and one of only a handful with the sign for 10,000.

At the bottom of the inscription, we can read three separate numerals: first 400,000 cattle (four "tadpole" signs below a bovine animal), then 1,422,000 goats (the god Ḥeḥ with arms extended upward to the right of the goat, with the other numerals below), and then, at the far right, 120,000 people. As described earlier, the Egyptian system is cumulative-additive; that is, it repeats the signs for the various powers of 10, up to nine times each, as needed, and the total value of the phrase is taken by simply adding up all the signs. There is considerable debate as to whether the numerals in the Narmer mace head should be interpreted as an amount of booty obtained by conquest, a sort of census or count of goods, or something else (Millet 1990). Because this is not a full hieroglyphic text—i.e., other than the *serekh* (royal seal) of Narmer, there are no phonetic signs whatsoever on the mace head—a final interpretation is elusive. What is not seriously in debate is that the numbers are surely inflated, particularly for the animals;

in other words, this is an exaggerated quantity that extols the excellence of Narmer in controlling such large quantities of humans and chattel. The mace head itself is one of a number of Early Dynastic display texts, rather than an administrative or archival one (Baines 1989). It existed to be seen at the temple, and the numerals are part of that display function. Our only evidence for large numerals (10,000 and above) in early Egypt comes from the mace head and other display texts. Allowing that the differential survival of materials and the incomplete archaeological record leave open the possibility of a more narrowly administrative numerical world in the early Egyptian state, these numerals were artificial and ideologized rather than meant to be true counts of anything.

Next, let us turn to Mesopotamia, and to the equally famous Weld-Blundell prism housed in the Ashmolean Museum in Oxford (AN1923.444), seen in figure 2.2. This is a clay "prism" roughly 20 × 9 × 9 cm in size and with cuneiform inscriptions on the four long sides in Sumerian, with a vertical hole through the entire prism to allow it to be mounted and rotated on a cylinder. The inscription itself dates to the Old Babylonian period, roughly 2000–1800 BCE, but the text shows evidence of being rather earlier in original composition and is written in Sumerian, which was an archaic and prestige language at the time.

The prism's feature of greatest interest (table 2.1) is the so-called "antediluvian Sumerian king list" which lists eight Early Dynastic rulers and their purported reigns (Young 1988; Friberg 2007). As with the Narmer mace head, the sheer magnitude of the numbers, well beyond a reasonable human lifespan, demonstrates their artificiality to a modern reader, and, frankly, probably to the Mesopotamian reader too. But these are not only artificially large numbers; they are also artificially *round* numbers in the sexagesimal or base-60 Sumerian numeral system. All but the last two are multiples of 3,600 (60 × 60), the second power of the base, and even those last two are multiples of 600 (60 × 10). It is a form of numerical play using the structure of the system itself. This contrasts with other, parallel king lists such as the ones in chapter 5 of the book of Genesis, where reigns such as Methuselah's 969 years are exaggerated but not otherwise *artificial*. In contrast to the Narmer mace head, the Weld-Blundell prism inscription is not really a display inscription—the cuneiform signs are tiny. But the function of the large numbers is similar: to overawe the reader at the massiveness of the span being considered.

Figure 2.2
Weld-Blundell prism, an Old Babylonian text in Sumerian cuneiform from Larsa, ca. 1800 BCE (source: Sumerian King List, 1800 BC, Larsa, Iraq by Gts-tg is licensed under CC BY-SA 4.0)

Table 2.1
Sumerian "antediluvian king list" on the Weld-Blundell prism

Name	Reign	Numeral expression	Transliteration
Alulim	28,800 years		8 × 3,600
Alalgar	36,000 years		10 × 3,600
Enmenluanna	43,200 years		12 × 3,600
Enmengalanna	28,800 years		8 × 3,600
Dumuzi	36,000 years		10 × 3,600
Ensipazianna	28,800 years		8 × 3,600
Enmeduranki	21,000 years		5 × 3,600 + 5 × 600
Ubartutu	18,600 years		5 × 3,600 + 1 × 600

Turning from the Old World to the New, figure 2.3 depicts Stela 5 from the Late Classic Maya site of Cobá on the Yucatan Peninsula (Graham 1997: 36; Gronemeyer 2004). Stela 5 is a block of limestone around 2.5 meters in height and bearing inscriptions on three sides. It expresses what is, by a substantial margin, the largest calendrical number in the Maya world. Normally, Maya dates are expressed through a series of five numerals from 0 to 19 in a vigesimal (base-20) sequence known as a Long Count, followed by a specific day name. Here, the standard Long Count date represented would be 13.0.0.0.0 4 Ajaw 8 Kumk'u; in other words, just the rightmost five digits of a 24-date sequence representing the "start date" of the Maya calendrical cycle, corresponding to August 11, 3114 BCE.

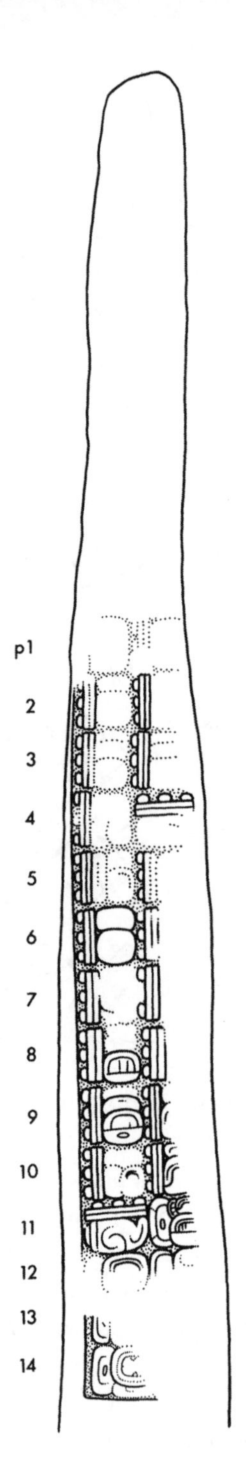

Figure 2.3

Late Classic Maya Stela 5 from Cobá, Yucatan, Mexico, drawing by Eric von Euw; © President and Fellows of Harvard College, Peabody Museum of Archaeology and Ethnology, PM 2004.15.6.18.9

But the actual date as written is 13.13.13.13.13.13.13.13.13.13.13.13.13.13.13.13.13.13.13.13.0.0.0.0. The series of thirteens (two bars for five and three dots for one) does not change the reckoning of the date—the "zero point" of the Maya calendar. But the extension of the calendar to 24 positions instead of five suggests to the viewer the ruler's capacity to imagine a universe of time extending far beyond the "zero point," over 28 octillion (2.8×10^{28}) years into the past—or one billion billion times the estimated lifespan of our universe since the Big Bang (Van Stone 2011)! The thirteens are conceptually like zeroes, and have been suggested to be read as such. It is as if we were to say "zero epochs, zero eons, zero eras, zero periods, zero millennia, zero centuries, zero decades, six years, three months, and two days"—the leading zeroes do not add to the quantity of time but they certainly add to its representation. This, and a small number of other Late Classic inscriptions known as "Grand Long Counts" that use the same calendrical conceit, are display texts intended to impress the reader with the vastness of time, and by extension, the power of the king to record and manipulate such quantities of time. Grand Long Counts are not "propaganda"—the date recorded is not false—but it is clearly ideological rather than serving some practical or historical purpose.

Perhaps no visual display of numerical magnitude is so impressive as the Roman *columna rostrata* (CIL 6.1300) honoring Gaius Duilius, consul and victorious admiral of the First Punic War (Kondratieff 2004). This inscription was first erected shortly after the events of 260 BCE that it records, but the text we have surviving today (figure 2.4) was almost certainly recut in the early Imperial period by Augustus or Tiberius.

Near the bottom of the text is an interesting passage (lines 14–16):

(ARGEN)TOM CAPTOM PRAEDA NVMEI ↈ
(OMNE) CAPTOM AES ↈ ↈ ↈ ↈ ↈ ↈ ↈ ↈ [—]
[—] ↈ ↈ ↈ ↈ ↈ ↈ ↈ ↈ ↈ ↈ ↈ ↈ ↈ [—]

The inscription, though fragmentary, can clearly be read as an account of Carthaginian loot plundered, in the amount of some millions of *aes* (a unit of currency), depending on how many signs are lost at the ends of the lines. Each sign composed of the strange concentric crescents represents a quantity of 100,000, the highest Roman numeral at the time of the Punic Wars. What we think of as the Roman M was originally an abstract sign, ↀ, which, when opened up from the bottom, comes to look like an M. The same sign but with two pairs of crescents represented 10,000, and then the sign we're looking at, with three crescents, represented 100,000.[2]

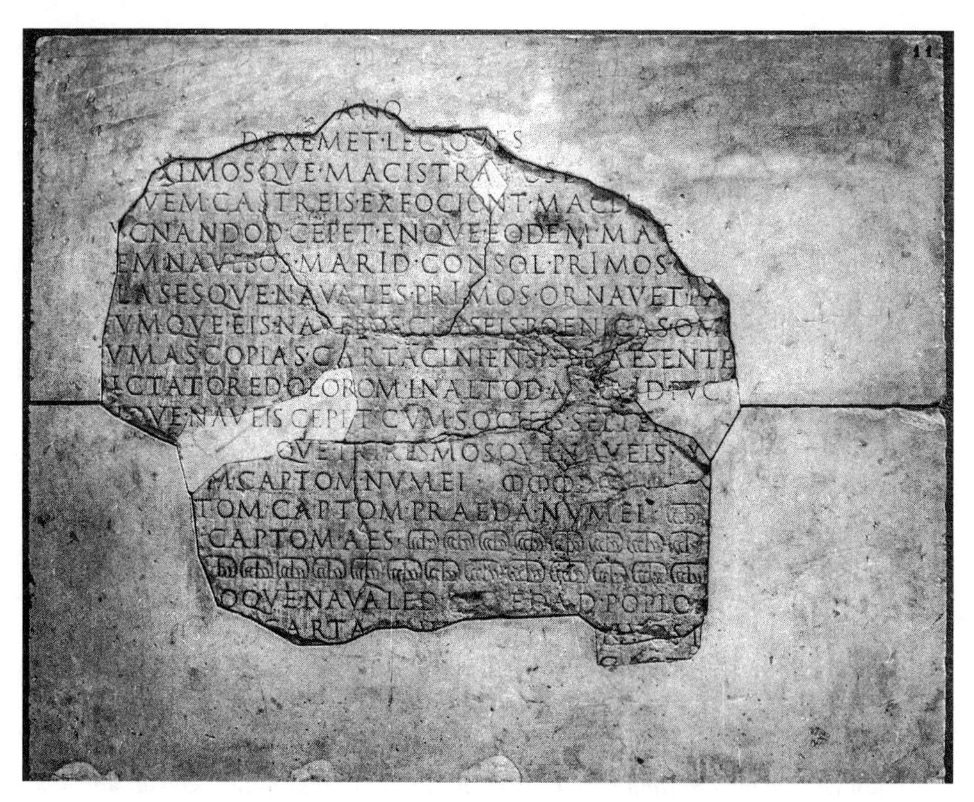

Figure 2.4
Columna rostrata of Gaius Duilius (CIL 6.1300; image courtesy of Center for Epigraphical and Palaeographical Studies, The Ohio State University)

Thus, the total amount is at least 2,200,000 *aes*, but Arthur Gordon, whose transcription I am following here, suggests that there could be as many as 15 missing signs (Gordon 1983: 125–126). Other than through repeating signs for 100,000, the other strategy available at the time of Gaius Duilius was to express the amount in words; the Latin expression (lacking a word for "million") for 100,000 would be *centena milia*, which could then be preceded by whatever numeral one wished. Later, by the first century CE, there would be other strategies possible; for instance, multiplication by 1,000 could be expressed by placing a horizontal bar (*vinculum* or *virgula*) above a numeral phrase, and by the late first century CE, multiplication by 100,000 could be expressed by placing two vertical bars and a horizontal bar on three sides of a phrase. But conciseness was hardly the goal here. In an inscription commemorating the quantity of plunder acquired, what better

Figure 2.5
National Debt Clock (National Debt Clock by Matthew Bisanz is licensed under CC BY-SA 3.0; source: Wikimedia Commons)

encomium than a massive text that visually imposes upon the reader—even, like most Romans at the time, a nonliterate one—the magnitude of the triumph?

By no means is this form of numerical display limited to the ancient world. I will turn at last to the philology of numerals in downtown Manhattan. If we were to travel to the southwest corner of 44th Street and 6th Avenue, near Times Square, we could see perhaps America's most glorious display inscription on the theme of fiscal responsibility: the National Debt Clock. Erected by Seymour Durst in 1989 and still operated by the Durst family of real estate developers, it impresses upon passersby the magnitude of the American national debt in large, constantly updating electronic figures.

Naturally, the number indicated on the clock is artificially precise; it represents some estimate of the national debt based on past rates of accumulation rather than an actual measurement of anything. But artificial precision is exactly what is desired here. If the number were 1,000,000,000,000 even, we could just write "1 trillion" or, using scientific notation, 1.0×10^{12}. That's what scientific notation is for—avoiding long strings of zeroes in very large or small numbers. And the artificial precision gives the audience—the passerby wandering along 6th Avenue oblivious to the financial troubles the Debt Clock is meant to emphasize—the impression that this ever-increasing number is constantly being calculated to the nearest dollar.

A minor crisis arose in 2008 when the national debt exceeded 10 trillion dollars for the first time and the display had insufficient digits, since at that time the leftmost electronic display showed an unchanging electronic dollar sign. Fortunately, after a brief interlude, the clock was reprogrammed and the electronic dollar sign was replaced with a shiny new electronic 1, which we see at left and which remains to this day (until it becomes a 2, soon enough). The dollar sign was then reappended awkwardly to the left of the first electronic digit as a sticker, a kind of billboard marginalia that, at the time the photo was taken, was already coming unglued at the bottom right corner. This workaround might, in a different context, be considered a failure—both of technology and of foresight. But I think it is better to understand it as a triumph of numerosity. The need to add another digit reinforced the magnitude of the very financial problem that the text was designed to highlight, and made national headlines. What better testament to fiscal profligacy (if that's your ideological inclination) than to run out of numbers?

What all of these texts, ancient and modern, share in common is that they employ what I will call *conspicuous computation*. Conspicuous computation is the intentional use of large numbers for their visual or psychological effect on the reader. This parallels directly Thorstein Veblen's concept of conspicuous consumption developed in his classic *The Theory of the Leisure Class* (1899). Veblen rightly noted that upwardly aspiring and well-off individuals in capitalist societies engage in conspicuous consumption, purchasing highly visible but unnecessary or even wasteful goods as a display of wealth. The prestige that accrues to them makes such practices rational, despite their seeming wastefulness. Similarly, conspicuous computation involves the production and display of numerical prowess far beyond that required for any purely representational or computational need. Not only may it require, in some cases, considerable effort to produce such numbers, but they are also cognitively expensive for readers to process: they are highly salient in size and require more time and effort to read. Conspicuous computation is widespread in hierarchical, inegalitarian state societies where numeracy is valued but unequally distributed among the populace. It may be more common in some societies than others—for instance, India is well known for the almost unimaginable vastness of quantities used in astronomical, calendrical, and ritual texts (Plofker 2009). Conspicuous computation is more than just numerical play (though it is surely, in some cases, playful). When used for display purposes, it serves as an index of the writer's control over numbers, and over the domains of whatever is being counted (time, cattle, money), and thus asserts status and claims cultural

capital. In short, numeration ceases to be a tool of arithmetic alo
serves instead principally as a textual tool of social control.

Now, large numbers can be spoken as well as written, and conspicuous computation (as I have defined it) might also include nonliterate as well as literate practices. Could spoken numerical language also work this way? Conspicuous computation reminds us of young children who, upon discovering the recursivity of the word *very*, become very, very, very, very, very annoying. In a previous study, I showed that the late nineteenth-century increase in education and, correspondingly, the ideological value attached to numbers led to a massive expansion in the English lexicon of *indefinite hyperbolic numerals*—number words like *zillion* and *jillion* that are used to signify enormous but otherwise unspecified quantities in everyday discourse, though not in formal contexts for the most part (Chrisomalis 2016). These numerals are highly useful in a milieu in which numerical reference is expected and normative, even where precision is either not possible or not desirable. They convey hyperbolic size in a pseudo-numerical fashion, without committing the speaker to any degree of specificity.

A pertinent analogy can also be drawn with several of the hierarchical but traditionally nonliterate societies of Polynesia, whose languages, such as Hawai'ian and Tongan, have seemingly fanciful large numbers for large powers of ten (100,000, 1,000,000, or even higher) as part of the numerical lexicon (Bender and Beller 2006). Bender and Beller contend that these numerical lexicons were not simply mental games, but served important social purposes in numerical practices—the collection and redistribution of resources, or the recounting of prestigious genealogies, for instance. It is in their public display and deployment that these numbers gain their power, so it does not really matter so much whether the numbers were accurate counts of anything, as they surely were sometimes but certainly not in others. What matters is the semiotic value of numerical expressions. As the linguistic anthropologist Judith Irvine (1989: 262) noted famously, "Linguistic phenomena are not all limitlessly and publicly available, like fruits on the trees of some linguistic Eden. Some of them are products of a social and sociolinguistic division of labor, and as such they may be exchanged against other products in the economy." Any speaker of Tongan might have been able to string together sequences of number words, but only those with the linguistic authority to do so in a way that was socially acceptable could accrue value from doing so.

However, there are also good reasons for us to separate spoken from written numerical practice. The first has to do with differences in modality, a concept I introduced in the previous chapter. Verbal numerals are impermanent whereas written numerals serve persistent display functions due to their relative permanency on linguistic landscapes. More significantly, in most written numeral systems, the physical size of a text often correlates with its numerical magnitude; "seven thousand" is longer than "two trillion" but 7,000 has four digits while 2,000,000,000 needs thirteen, in Western numerals. The cognitive cost of processing those digits is significant; so, for instance, considerable effort was needed to count the signs for 100,000 on the *columna rostrata,* or all those signs for 13 on Stela 5 from Cobá, especially given that there was no textual technique used to break up digits as we do with a comma after every third digit. These sorts of conspicuous texts take time and energy to read, even as they draw the eye immediately. They are both highly salient at first glance, and then highly challenging to read afterward. Houston and Stauder (2020: 26–27) highlight a similar phenomenon for the Egyptian and Maya hieroglyphic writing systems, which are characterized by visual excess in the complexity of virtuosic sign-forms meant to draw in readers and demand their attention. Conspicuous computation similarly builds on the limitations of the human mind to process complex visual imagery.

But not all instances of conspicuous computation are so visually distinct. The signs of the Sumerian king list on the Weld-Blundell prism are tiny and short, even as the amounts they express are huge. Here the semiotic effect is more indirect; because administrative and bureaucratic practices use numbers for recordkeeping, long-distance communication, and sometimes directly for calculation, written numerals serve as an index of state power. Because numerical notation was principally developed in and controlled by states in the premodern world (and this is true even today), baffling one's readers with the enormity of some quantity serves an important symbolic function. This holds true even when, as with the National Debt Clock, the argument being advanced is a negative one—only a very powerful state indeed could be trillions of dollars in debt in the first place.

Finally, while full phonetic literacy was rare, the ability, at some level, to draw symbolic inferences from written numerals is much more widespread. You do not need to have full control of the Maya calendar to know that bars are five and dots are one, and that the number expressed on Stela 5 at Cobá must be enormous. Written numerals sit at a tipping point: they are just comprehensible enough to be used in display inscriptions and be understood by the largely nonliterate, and just incomprehensible enough (especially in their use) to mystify. To a large extent this is true even in our hyperliterate era. As John Allen Paulos notes, while illiteracy is seen as a sign of stupidity, some people seem almost proud of having no proficiency with numbers (Paulos 1988).

In conspicuous computation, we see one of the challenges to Zipf's dynamic philology. Brevity may be the soul of wit, but magnitude is the heart of power. Just as we recognize that sesquipedalian words—i.e., following Horace (*Ars poetica,* 97), literally *sesquipedalia verba,* "words a foot and a half long")—serve as tools of social distinction and an index of education, conspicuous computation draws the reader's attention to the intellectual and social capital needed to produce and manipulate large quantities. For most humans—elite and nonelite alike—large numbers above around 1,000 are not understood concretely or specifically but as vast, vaguely indeterminate quantities. Landy, Silbert, and Goldin (2013) show that when American participants in an experiment were asked to place the numbers *1 thousand, 1 million,* and *1 billion* on a number line, around half of them put *1 million* halfway between the other two, even though one million is much closer to one thousand than to one billion. The principle of least effort can thus be consciously violated as part of a political strategy to impress.

This is not at all to say that Zipf was wrong, but rather that the exception proves the rule (mindful, once again, that "prove" here is akin to "probe"). We see that Zipf's principle generally holds true because of the power that accrues from its conscious violation. In the same way, the archaeologist Bruce Trigger noted that in early states, large-scale monumental architecture served as a powerful demonstration of symbolic authority, as a flagrant violation of the principle of least effort, through the seemingly wasteful expenditure of surplus labor and goods (Trigger 1990). Building an imposing pyramid or a glorious wall, when one does not really need to do so, is meant to reinforce the power of a leader to command labor and resources. Conspicuous computation requires still less, in that its energetic requirements are low relative to the benefit obtained through display inscriptions. A conspicuous computation is like a picture of a pyramid or a blueprint of a wall, but used in a context where its visibility aims at the same effect.

Roman numerals redux

This detour into the world of massive numerical inscriptions, while seemingly far afield from the decline of the Roman numerals, brings us back to Zipfian questions about frequency and about effort. I hope to show that these same sorts of comparative and cognitive approaches can be applied fruitfully to the large-scale and diachronic history of numeral systems. I want to start with the presumption that the Roman numerals, having survived for two thousand years, during most of which the dozens of other European and Mediterranean numeral systems posed them no significant threat, must have been good for something. The Romans were not dupes.

Let's start with the problem of conciseness—a key Zipfian question. It is indisputable that the number 1,888, written as MDCCCLXXXVIII, is extremely long: four digits compared to 13. This seems bad. But also note that 2,000 has four digits in Western numerals but only two (MM) in Roman numerals. If one were to take the mean of all the integers under 1,000, you would find that the Western numerals average 2.89 characters per number while the Roman numerals are up around 8. (I will address this issue of conciseness further in the next chapter.) All things being equal, the Roman numerals are longer—but all things are not equal. Note that the numbers where the Roman numerals are shorter than Western numerals are the round numbers: 10, 50, 100, 500, and 1,000 each only take one sign, for instance. And it is a remarkable property that in any reasonably large corpus of natural language, round numbers are more common than nearby nonround numbers (Sigurd 1988; Coupland 2011). In any substantial body of texts in any decimal language, the number 100 will be more common than 99 or 101. This may be because humans are lazy (and prefer to write shorter numbers), or because, pragmatically speaking, precision is not always valued.

In his famous Maxim of Quantity, the philosopher of language Paul Grice (1991) expresses the principle that from a pragmatic point of view, the speaker or writer should be as informative as they need to be, and no more. When pragmatic maxims are violated, the job of the analyst of language is to look for the meaning underlying the usage in question, as a means of approaching the intention of the speaker or writer. Some numerical examples may be pertinent. When the pedant insists that 2001, not 2000, started the new millennium, he is indexing himself as the sort of person who has a particular sort of knowledge about the world, which he is placing on display for the (presumably ignorant) hearer. When a child says that she is "seven and seven-twelfths" years old, her seemingly excessive precision violates the Maxim of Quantity in ordinary circumstances, but shows the hearer that for her, it matters quite a lot that she is no longer just "seven" or "seven and a half." She may also be demonstrating a new fluency with fractional and calendrical thinking that is precocious for her age. Conspicuous computation fits well within this framework from linguistic pragmatics, as a flagrant violation of the Maxim of Quantity.

In most cases, the Roman numerals obey Zipf's law of least effort: the most frequent numerals—the low numerals, and the higher round numerals—are shorter than the rarer ones. In contrast, for the Western numerals, while conciseness holds for the low numbers, round numbers enjoy no such advantage. Overall the Western numerals are still shorter, but the Roman numerals make up for this deficit in ordinary, unmarked contexts because most numbers are not really that long to write, and round numbers are especially short. In contrast, when you need a big number—not just one that has a large quantity, but a big number, such as when you need to inscribe a date on the lintel of a majestic building, or on the frontispiece of an old-fashioned book—the Roman numerals are there for you. Make the reader work a little—both because the inscription is big, and also because we aren't as familiar with Roman numerals as we used to be.

An interesting example of this principle at work comes from one of the contexts where Roman numerals are still employed regularly, the numbering of Super Bowls, the annual championship of American football. For years, this was the tradition, accruing all the prestige and majesty of the older numeral system that modern Western societies also give to royalty. Even as Super Bowl XXX wreaked havoc with the early online search engines of the 1990s, flagging the event as adult content, the practice continued. And Super Bowl XL a decade later not only signified the number 40, but also indexed an extra-large event—a day of great import. But in late 2015, word came down from the NFL that the tradition would be suspended—that the 2016 event would be "Super Bowl 50," so as to more prominently feature the numeral 50. The symbolic association of L with loss in sporting events surely played a part in avoiding "Super Bowl L," just as Western buildings often lack a thirteenth floor, or East Asian ones a fourth floor, due to the inauspicious associations of those numbers. But another factor was that, for the first time in forty years, we were faced with a tiny Roman numeral, too concise to serve its symbolic purpose. And the X of Super Bowl X, back in 1976, is surely more symbolically potent, not to mention more symmetrical, than the lowly L. And so the change was made, though not without some wailing from traditionalists. But sure enough, Super Bowl 50 was followed by Super Bowl LI, the use of Roman numerals resuming in 2017 once the crisis of insufficiently conspicuous computation had passed.

The Roman numerals share with a whole class of cumulative numerical systems (like the Sumerian and Egyptian systems discussed above) the property of repetition of identical signs. Cumulation is very useful for the writer interested in conspicuous computation, because even rather small numbers may end up having a considerable graphic size. Because tallying—the use of cumulation without a numerical base to take an ongoing count—is attested as early as the Upper Paleolithic, it has been part of human notational practice for over 20,000 years and likely longer (Marshack 1972; d'Errico and Cacho 1994). Cumulative structures need not be conspicuous, but they afford conspicuous computation.

Having said this, cumulation has its limits, and the desire to express size through notation is counteracted by the sorts of considerations of efficiency and effort that motivate Zipf's law in language. Frequently, up to three or four signs are grouped together in a single register or cluster, but beyond that point, some way of dividing them up is used—for instance, when tallying, by putting a horizontal stroke through four to delineate groups of five, or with Roman numerals, by using a subbase of 5; or with Egyptian hieroglyphs, by putting like signs in groups of no more than four signs in separate registers, or with Phoenician numerals, by simply breaking up lines of five or more signs into groups of at most three, using spacing (table 2.2).

Table 2.2
Tallying strategies

System	"7"	Subitizing enabled through
Unstructured	IIIIIII	None
Tallying	~~IIII~~ II	Stroke directionality
Roman numerals	VII	Subbase
Egyptian hieroglyphs	IIII III	Register organization
Phoenician numerals	III III I	Horizontal spacing

The reason that all these otherwise quite different notations break numbers up in this way is because of a capacity called *subitizing*, from the Latin *subitus* "sudden." For up to around four items, humans (and many other species, in fact) are capable of rapidly assessing the quantity of some identical objects, without needing to count them individually (Kaufman et al. 1949; Mandler and Shebo 1982). So, for instance, if I were to toss three quarters on the floor at a lecture, I presume that several audience members would immediately make a dive for them, but just before they did, they would rapidly and without counting perceive their numerosity as three (at least, if the quarters landed fairly close together). Subitizing is an aspect of our capacity for numerosity, but it is also a basic part of our capacity for object recognition. Beyond the limit of around four, most nonhuman animals are incapable of exact quantification, while humans normally deploy a counting strategy, which is slower and trickier. Language also plays a role here—while subitizing appears to be language-independent (as one would expect for a capacity we share with many nonhuman animals), exceeding the limit of three or four is made easier by having a set of number words in a sequence, which not all human groups have found much need for (Overmann 2015). If I were to throw seven quarters on the floor, while the audience might dive just as quickly to retrieve the bounty, the gathered crowd would need to count them to know the quantity.

Cumulative numerical notations, as well as tally systems that use cumulation, are extremely common cross-culturally, and almost all use this constraint on human cognition to subdivide larger bundles of like signs, because unstructured groups, like the first row of the table above, are very difficult to read once the number of items gets above the subitizing limit. In contrast, systems like the Egyptian hieroglyphic numerals and the Roman numerals are not particularly difficult to read, with a little training, even when we consider the occasional (and in classical texts, rather sporadic) use of subtraction in numerals like IV and XC. The subitizing limit also plays a role in ciphered systems like the Western numerals; most style guides allow you to write 1000 (although 1,000 is also common), but once you hit a five-digit number like 10,000, the division of numerals into groups of three using commas increases readability—and also corresponds with the structure of the English lexical numerals. These textual strategies are very useful ways of breaking up numbers that would otherwise be extremely unwieldy for the reader. And yet sometimes, as in the *columna rostrata* with its 22 or more signs for 100,000, strung together, breaking the signs up would defeat the purpose.

A final example of conspicuous computation makes use of these features of cumulative notation as well as sign grouping. Hanakapiai Beach on the island of Kauai, Hawaii, is an extraordinarily beautiful but also dangerous beach, frequented by hikers and those tourists looking to get away from the traditional tourist traps. Because of the dangers of rip currents, a sort of informal yet very potent carved wooden sign has been erected as a warning to visitors (figure 2.6).

Figure 2.6
Tally marking at Hanakapiai Beach, Kauai, Hawaii (Hanakapiai Beach Warning Sign by God of War is licensed under CC BY-SA 3.0; source: Wikimedia Commons)

Toward the bottom of the text, sixteen groups of five tally marks, plus two solitary marks beyond that, indicate the number, 82, of people who have purportedly met their end due to Hanakapiai's powerful currents. But note that while this text is rough-hewn, it does not appear to be really an ongoing count—it's not as if the marks are of different lengths or seem to have been cut by different hands. Rather, carving eighty-two separate marks, organized in fives for a little readability, but still requiring that the reader count, laboriously, sixteen groups plus the two stragglers, uses an enormous amount of textual space compared to simply writing 82. The organization of tally marks by fives deals with the subitizing limit, but only for each group; because there are sixteen groups of five, the reader still needs to count the groups. The text draws the eye, forcing the reader's cognitive attention on the scope of the problem and the effort required. Because the text has the appearance of an ongoing tally, rather than a completed count, the writer has also very cleverly left a wide space at the end of the count, as if to suggest that the reader should beware of having their fate notated with the eighty-third stroke. This is, then, a very clever text, combining the cumulation of 82 signs to produce an emotional effect in the reader, the grouping of tallies into chunks to facilitate taking the count, and the open-ended use of space to further highlight future dangers.

Conclusion

The sorts of analyses of individual exemplars of conspicuous computation that I have outlined in this chapter have much more in common with classical philology as traditionally conceptualized—as a way of analyzing and interpreting texts in relation to one another, and in relation to writerly intentions and reading practices—than they do with the sorts of large-scale quantitative philological analyses advocated by Zipf. And that's fine—much as we would not all want to wear the same clothes, or use the same languages, we can allow a disciplinary space for a broad philology—and by extension, for a broad anthropology—where multiple methods and perspectives flourish in parallel, intersecting only in these strange liminal spaces. Even though the methodology employed here is different from Zipf's, it converges to similar conclusions. That's because the texts described above are exactly the sort of exceptions that prove the rule—that show us why Zipfian principles like least effort are generally true.

Another such space is the subdiscipline of cognitive philology, which uses approaches from cognitive science to analyze literary texts. So, for instance, Gianluca Valenti (2009), noting that the human capacity to memorize information is limited to around seven (plus or minus two) elements, following the classic findings of George Miller (1956), argues that medieval European poets made decisions about prosody and meter that respected these cognitive limits, so long as poetry was largely transmitted orally rather than through writing. The method Valenti employed—analyzing what medieval poets wrote about the aesthetic canons of their genre—seems very far from cognitive science, but the principle being used is taken right from its core. The argument I have raised above similarly acknowledges the existence of these sorts of cognitive constraints, but then forces us to attend to those cases where they are flagrantly violated, to great emotional and cognitive effect.

I suggested earlier that philology had good reason not to see much use in Zipf, just as he saw little use in his home discipline. A universalist, positivist, and synchronic philology has little to hang onto, in terms of method or theory, that would appeal to the historical disciplines. But the solution is not a retreat into particularism and disciplinary silos—to do so would deny that we have anything to say to one another. Instead, I think we need a new triad. In place of a universalism that assumes that there will be massive regularity or even cultural laws, we ought to be engaged in a judicious comparativism. Comparativism takes as a working assumption the idea that cross-cultural patterning is real and meaningful and worth looking for, without assuming that patterns need be exceptionless. Actually, it is in the exceptions that we are able to determine empirically why the patterns might exist, and the conditions under which they might not pertain. In place of Zipf's rigid positivism, but also in opposition to a purely hermeneutic epistemology, I think we ought to look to cognitive principles. At least for the domain of number, attention to the general cognitive properties of numerical cognition, as well as the biases and schemata through which knowledge systems are created, can be a productive strategy. Because cognition takes ideas and the mind seriously, it gets past behaviorist pitfalls while nonetheless insisting that there are meaningful integrations for philology with disciplines like psychology, linguistics, and anthropology, where such ideas are already bearing fruit. Finally, to complement a synchronic approach, a dynamic philology requires diachronic analyses, including data that allow us to study change through time. By focusing on events and processes rather than just outcomes, we open ourselves up to a richer set of influences from both the humanities and the sciences.

Understanding the role of attention, memory, and effort in processing numbers is absolutely critical if we are to address the purported problems with the Roman numerals. To what extent was their decline brought on by their inefficiency? I have already suggested that, far from being inefficient, they clearly were useful for many writers and readers for many years. If the principle of least effort were the main one in operation, surely the Roman numerals would not have survived for two millennia, essentially unchanged. To resolve this issue, we need both to look at the actual properties of the system itself, to look at its competitors (chiefly the Western numerals), and then, most importantly, to confront the historical factors surrounding its centuries-long decline in frequency.

3 / III The decline and fall of the Roman numerals, I: Of screws and hammers

Why don't we use Roman numerals anymore? Over the past twenty years, I've asked this question to audiences ranging from schoolchildren to mathematicians. The answers I receive, though they differ in structure and complexity, echo a single theme of computational efficiency. Faced with the superior Western numerals, the cumbersome Roman numerals were unable to compete, and so they faded out of everyday use, retained only for prestige purposes where their defects are overcome by their archaic majesty. Or so the story goes.

But how can we empirically investigate whether this was actually the case? How can we establish under what circumstances, and with what motivations, the users of Roman numerals between the thirteenth and eighteenth centuries shifted from one notation to another? That the shift happened is certain. Almost all of the hundred or more numerical notations used historically are no longer used today. Extinction, or at least obsolescence, seems to be a natural part of the life cycle of number systems. At present, the Western numerals 0123456789 are more or less ubiquitous. The question, then, is not just about the Roman numerals, but about what motivates people to abandon one way of writing numbers in favor of another. Are people actually motivated principally by issues of efficiency—and what does "efficiency" mean? And if not, what else can explain these events of abandonment and replacement?

The Roman numerals provide a good case study because they are, surely, the best-attested case of a once-popular numerical notation falling out of everyday use. They were developed on a Greek and/or Etruscan model around the fifth century BCE and, by virtue of their association with the Roman republic, became predominant in the Italian peninsula. From around the third century BCE to the fifteenth century CE, most of western and central Europe used them for purposes both mundane and lofty. During late republican and imperial Rome, their geographic scope was still broader, into much of North Africa and the Middle East. Most of Europe used them throughout the Middle Ages. But starting around 1200, and accelerating rapidly after 1500, they were abandoned, sometimes rapidly, more often haltingly, until by 1700 they were, if not a curiosity, definitively archaic. Today we retain them only for secondary functions (numbering the pages in the prefaces of books, for instance) and to count things to give them an air of prestige (kings, popes, Super Bowls). The transitional period, focusing on Western Europe in the late medieval and early modern period, is exceptionally well-documented, and we have a wealth of different kinds of evidence showing the rate and context of the Roman numerals' decline.

It is not implausible that the structure of a notation would be relevant to whether it is retained or abandoned, so we need to look at any cognitive factors that could be employed to compare the representational properties of different numbering systems. At the same time, notations, writing systems, and mathematical and scientific practices do not operate in a vacuum, so we should also examine the social, political, and economic factors that we know to operate on the transmission and adoption of new technologies. We need to be wary, lest we fall into the same sort of logic employed by racist anthropologists and linguists who saw languages that went extinct as a natural byproduct of the inferiority of their speakers. If we have a null hypothesis, it should be that considerations of efficiency did not lead to the replacement of the Roman numerals, until there is firm evidence that it did.

In order to establish whether representational effects of numerical systems motivated a systemic transformation, we need to address four key questions:

1. Are there representational effects of numerical notation systems that we can identify as providing cognitive advantages?
2. Were those advantages relevant to tasks and contexts for which numerals were used in a specific society?
3. Were the notation's users aware of those advantages, and did they perceive them as being relevant to their needs?
4. Were those advantages effective in producing a transformation or replacement of a numerical system?

If we can answer yes to all four of these questions, then the traditional view of the decline and fall of the Roman numerals holds. If not, then things get more complicated, and more interesting. We are in a sort of mystery novel in reverse: "whodunit" has been answered many times already, almost in unison, at the beginning. The Western numerals are the guilty party. But by the big reveal in the final act, we'll see that the when, how, and why are very confounding indeed.

Evaluating the merits of numerical notation

The scholarship on mathematical notations is replete with commentary asserting that the cumbersome nature of the Roman numerals caused their eventual decline and replacement, and while we shouldn't take these comments at face value, we also don't want to ignore all the modern observers who are unanimous in this regard. But we should be cautious. It is very difficult for anyone to evaluate the utility of a system without considering the social context in which systems are learned and used. Very few of these commentators actually provide direct evidence for their suppositions. So why do they believe them?

Most school curricula in North America and Western Europe, at some point, introduce children to Roman numerals. After all, they are common enough still that not knowing them would be a disadvantage—they form a part of the cultural capital that most Western societies expect their members to acquire as part of a good education. However, by the time that children are old enough to be taught to use Roman numerals, they are immersed in a mathematical world where the Western numerals and the principle of place value are utterly ubiquitous. What we learn as children about Roman numerals is also far removed from the way anyone actually learned Roman numerals in the past. As Durham (1992: 28) notes, "It is today an occasional source of amusement for arithmetic teachers to threaten children with having to do 'long division' in Roman numerals." Long division, as we understand it, is a product of sixteenth-century mathematics; it was designed to be used with the Western numerals 0–9 and place value, and has nothing to do with Roman numerals. A cynic might suspect that teaching the Roman numerals this way was designed to get students to think they're useless.

We might ask, then, how else we might know what the cognitive or structural properties of the Roman numerals are. Cognitive psychology and cognitive neuroscience are hard to use for this purpose, because no other numerical system, other than the Western numerals, is used and taught widely in schools anywhere. Early work by the cognitive scientists Gonzalez and Kolers (1982) and Noël and Seron (1992) involved presenting Roman numerals and Western numerals to Western students. Both studies used only low numerals (up to 12) and investigated how quickly students could read them, finding in each case that Roman numerals were read more slowly. But this shouldn't surprise us: familiarity is predictably related to speed in all sorts of contexts. Interestingly, though, Noël and Seron (1992) found that while the students were slower to respond regardless of which numeral they were presented, for the numbers 1 (I), 2 (II), 3 (III), 5 (V), and 10 (X) the difference was not as great. This suggests that there is a notational effect, but that it depends on how specific numbers are encoded. We might predict, for instance, that 300 (CCC) would be read more quickly than 65 (LXV) because the former has three identical signs while the latter has three different signs. If descending sign order is relevant, we might also predict that XVI (16), which is written with the numerals in descending order, only using addition, would be read more quickly than XIV (14), which uses subtraction. However, this area of research has never been extensively followed up, although it readily could be.

In any case, a challenge to the validity of the general conclusion that the Roman numerals are slower to work with is that the Roman numerals are going to be less familiar for almost everyone, and it is not obvious that more training can resolve this difficulty—or what such training would even entail. Probably this is why there are so few cognitive studies of the use of Roman numerals—such studies are all, to some extent, artificial. For verbal language, in contrast, differences in morphosyntax among the world's numeral systems can be evaluated using cognitive-psychological methods experimentally, because there are plenty of fluently bilingual people, and lots of variation even among monolinguals.

Moeller et al. (2015) studied the effect of direction of writing along with the ordering of numbers among speakers of Arabic, English, German, and Hebrew in a fascinating experiment. They found that speakers of languages where ones precede the tens in verbal numerals, such as German, where 47 is *siebenundvierzig* "seven and forty," experienced interference across modalities in reading digits as number words—but only where the script was written from left to right. Hebrew speakers, who read number words right to left in descending (tens-units) order, but Western number signs left to right, also experienced interference. Arabic speakers, who have "units-and-tens" numeral words written right to left in the Arabic script, but who read numerical notation left to right, had no such interference, however, nor did English speakers, because these two languages have consonant directions of reading both modalities. We can thus see that notation does have an effect when combined with the order of words in a language and the direction of its ordinary writing.

If we want to compare numerical notations to one another, though, this sort of comparison is not a realistic option—almost every literate person is a native "speaker" of Western numerals and of no other system. The direct cognitive approach is suggestive but won't show us how historical users of the system perceived and used the Roman numerals. We can, however, evaluate the properties of numerical notations in the abstract that might be relevant cognitively. Three properties of numerical notations that we might look at and evaluate are *conciseness*, *sign count*, and *extent* (Chrisomalis 2010: 389–399), even in the absence of any knowledge of how they were used. Bender, Schlimm, and Beller (2015) use a similar approach, though with some differences, in their important cognitive study of Polynesian numeral words. They compare the numeral word systems of different Polynesian languages, paying specific attention to specialized systems used for counting only one sort of object (like coconuts). They show that the properties of these systems are useful in contexts where mental arithmetic, rather than written arithmetic, is used to hold in memory and work with large numbers of objects beyond the limits of the ordinary number system.

Conciseness is the length of a numeral phrase in a particular notation. The conciseness of a particular expression is relatively unambiguous: the length of the Western numeral 2,016 is 4 and of the Roman numeral 2,016 (MMXVI) is 5. To evaluate the *systemic conciseness* of an entire numerical notation system as a whole, however, is more complicated. It requires, firstly, that we establish some arbitrary basis for comparison, such as measuring the average conciseness of numeral phrases from 1 to 1,000. The results would be different if the comparison scale chosen were different. It also compels us to disregard variation in particular numerical expressions—for instance, should we use the Roman numeral IX or VIIII as the relevant basis for comparison with 9? Even to this day, Roman numeral clocks normally read IIII for 4, not IV. It has been shown experimentally that when asked to draw a Roman numeral clock from memory, people typically, and erroneously, write IV, probably because they have learned that that is the "correct" expression in school or in other contexts (French and Richards 1993; Richards 1996).[1] These studies show that the cognitive schema of the Roman numerals acquired in non-clock contexts overrides the memory of actual Roman numeral clocks. What "counts," in this case, as the real Roman numeral for 4?

We should thus be evaluating conciseness not only in the abstract but controlling for the relative frequency of particular numerical expressions in each written tradition. So, for instance, the number 1,888 is far longer in Roman numerals than in Western numerals (MDCCCLXXXVIII = 13 signs) but it is really quite rare. Conversely, the number 10 (X) is shorter in Roman numerals than in Western numerals, and it is exceedingly common. As I discussed in chapter 2, for many round numbers, Roman numerals are equally or more concise than Western numerals. But as we saw, round numbers (powers of the base, and multiples of the powers of the base) are among the most frequent numbers in written texts (Coupland 2011). To evaluate the conciseness of a system appropriately, we would need a text corpus including numerals, and would need to take the average conciseness of numerals as actually written, to control for this factor.

Bender, Schlimm, and Beller (2015) argue for the cognitive advantages of domain-specific specialized counting systems in three Polynesian languages, Tahitian, Mangarevan, and Pohnpeian. All three of these languages have two number systems: a general-purpose one, and a specialized system for counting coconuts. They evaluate the representational compactness of number systems (how many morphemes it takes to express each particular number), and from there aggregate these to compare different systems within a single language family. They show that Polynesian object-specific coconut-counting systems have advantages in compactness over not only the ordinary number word systems of those languages but also those of languages like English. They argue persuasively that these object-specific systems, sometimes seen as concrete, cumbersome, or unnecessary, have cognitive advantages. Because they are not working with natural language data, they cannot account for the differences in relative frequency of different number words, however.

Sign count is the inventory of numerical signs needed to be known for a user to competently represent any number. For the Western numerals, the sign count is 10; for the Egyptian hieroglyphs, which have one sign for each power of 10 up to 1 million, the sign count is 7. Again, what looks straightforward at first glance becomes more complicated. For instance, do we need to count the decimal point as a numeral sign, or the comma that goes between chunks of three digits larger than 1,000? What about variant paleographic forms of numerals for 2 (with a loop), 4 (with an open top), and 7 (with a line across the diagonal), which you might not write, but certainly need to know in order to be fully competent at reading Western numerals? What about signs (such as the Roman numerals for 10,000 and 100,000) which were used only rarely, by a few authors, or at certain periods?

In general, we can presume that needing to memorize fewer signs imposes a lighter cognitive load than needing to memorize many. Funes (in chapter 1), the memorious, could have a different word for each number, but we do not have this luxury. But there are also tricks we can use to mitigate the effect of a large sign inventory. For instance, many numeral systems use the letters of the alphabet, in their preexisting order, as numeral signs. A Russian user of Cyrillic numerals only would have needed to know 27 letters and their order, corresponding to 1–9, 10–90, and 100–900, where an English user of Western numerals needs to know the 26 letters of the Roman alphabet *plus* the 10 digits of the Western numerals. Which is more convenient—and how would we know?

Finally, we might ask how high a numeral system extends. The linguist Joseph Greenberg (1978) defined the limit L of any lexical numeral system as one greater than the largest number expressible in a particular system; Bender, Schlimm, and Beller (2015) call this same concept *extent*, and I will follow them here. Extent can be operationalized as the largest number not normally expressible in a numerical notation system without developing new signs. Every lexical numeral system has some limit—there is always some inexpressible number in a system, beyond which you need a new word. But for numerical notation, positional systems like the Western numerals are almost always infinite in extent; in theory, we can just keep on adding zeroes to a number to multiply successively by ten.[2] The extent for noninfinite systems generally depends on the highest sign for some power of the base—so, for instance, the Linear B numerals used in Mycenaean society have signs for 1, 10, 100, 1,000, and 10,000, each of which can be repeated nine times, so this system's extent would be 100,000.

Extent is relevant, however, only insofar as the numerical needs of a particular society require the notation of large numbers. Users of numerical notations have proven to be very effective at creating new symbols and principles when needed for specific numerical functions—necessity is the mother of invention. As a practical matter, extent is not always considered valuable or important, especially where it interferes with conciseness. This is true even of systems with infinite extent. English-speaking writers would almost always write 100 quadrillion rather than 100,000,000,000,000,000 (one followed by seventeen zeroes).[3] In chapter 6, I show that because numerical expressions mix modalities in this way often (e.g., mixing number words with symbols), we need to compare not just words versus notations, or one notation versus another, but also cases where hybrids are advantageous.

The counterargument could be raised that a truly infinite numerical notation system is the only one that has a representational effect allowing people to develop the concept of infinity. This is a sort of claim that using a particular numerical notation causes a specific cognitive effect; in the same way, writers such as Peter Damerow (1996) have argued that, in Mesopotamia, changes in notation facilitated abstract rather than concrete numerical thought. But I could readily make the argument the other way: that it is only once the infinity of the numbers is conceived that the need to notate such staggeringly large numbers arises. For instance, in Indian literature, in the *bhuta-sankhya* system of "object numerals" words had numerical values from 0 through 9, and could be concatenated in metrical verse to represent numbers in a way resembling place value (Plofker 2009: 47). But this system was used as early as the third century CE, at least three hundred years before the Indian numerical notation started to use place value and a special sign for zero. The Sanskrit concept of *sunya* "emptiness, void" was repurposed as a word for zero, and then, arguably, as a sign for zero. Only in the sixth century do we find good evidence for place value numerical notation. Still later, in medieval Indian mathematics, new number words were developed for almost unimaginable quantities, the largest attested of which appears to be the Sanskrit term *asaṃkhyeya*, which could be as large as 1 followed by 74,436,000,000,000,000,000,000,000,000,000 zeroes, depending on which translation you follow (Yong 2008). These numbers are not infinite, but are far larger than any quantity that any human has actually needed for any practical purpose.

In summary, we can see that any of conciseness, sign count, or extent might be relevant to the historical changes among the world's numeral systems. But how would we know which is most important, if any? No system is ideal in all respects. Ciphered-positional systems—those structured like the Western and Arabic numerals—are less concise than ciphered-additive systems like the Greek, Hebrew, or Cyrillic alphabetic numerals.

For any Western numeral that contains zeroes, an alphabetic system has none—where we would write 2,019 with four signs, it would need only three signs, for 2,000, 10, and 9 (in Greek alphabetic numerals, ͵βιθ). Similarly, ciphered-positional systems have a larger sign count than cumulative systems—ones that depend on the repetition of signs, like the Roman numerals. How do we gauge what mattered to past people? To summarize, table 3.1 compares otherwise generic, base-10 systems, with no other irregularities, across the five basic types outlined in chapter 1.

There are certainly other features that could be considered relevant as well—for instance, we might try to operationalize irregularities in systems such as subtractive notation (XIX instead of XVIIII for 19 in Roman numerals) or irregular ordering (the reversal of the units and decades in several alphabetic numeral systems when writing 11–19). We might also develop an index for correlating the structure of a numerical notation to the structure of the lexical numerals of speakers of various languages, as a means of determining whether this aids in cross-cultural transmission.

One potentially important factor on which ciphered-positional systems really do seem to have an advantage over other systems is one not previously discussed in the literature on numeration. I will call this *size ordering:* the property comparing the length of a numeral phrase with its size. So, for instance, note that all the two-digit Western numerals are larger than all the one-digit ones, and all the three-digit numerals are larger than the two-digit and one-digit ones. In other words, it is possible, simply by looking at the length of a numeral phrase, to get a rough idea of its numerical magnitude. You do not need to know anything else to know that 2,000 is bigger than 20. In contrast, the Roman numerals MM and XX have the same length, and a low number like 38 (XXXVIII) is much longer than either of them, so they

Table 3.1
Conciseness, sign count, and extent (after Chrisomalis 2010: 397)

	Conciseness	Sign count	Extent
Ciphered-additive	2.70 (1)	18–30 (5)	Normally 10,000–1 million (4)
Ciphered-positional	2.89 (2)	10–11 (3)	Normally infinite (1)
Multiplicative-additive	4.49 (3)	12–14 (4)	Normally 100,000+ (3)
Cumulative-additive	13.59 (4)	4–7 (2)	Normally 1,000–100,000 (5)
Cumulative-positional	13.78 (5)	1–3 (1)	Normally infinite (1)

Note: Numbers in parentheses rank the five types in terms of each of these features.

are not well size-ordered. Figure 3.1 compares the size ordering of four well-known numerical notations for all numbers from 1 to 1,000: the cumulative-additive Egyptian hieroglyphic numerals (base 10); the cumulative-additive Roman numerals (base 10, subbase 5); the multiplicative-additive classical Chinese numerals (base 10); and the ciphered-additive Greek alphabetic numerals (base 10). The Egyptian system (the top, black line) is the least concise, and it also experiences the wildest fluctuations in length, culminating at the right side of the graph where 999 has 27 signs but 1,000 only has 1. The Roman numerals, by adding a subbase of 5, mitigate this effect somewhat, and the Chinese numerals more so. Even though the Greek alphabetic numerals are more concise than the Western numerals, they have these differences in size ordering. If we were to add the Western numerals to this graph, they would simply proceed upward stairwise, never decreasing.

This property of numerical notations has not attracted any significant attention. Is it relevant? Well, one reason to suppose it's not is that no one

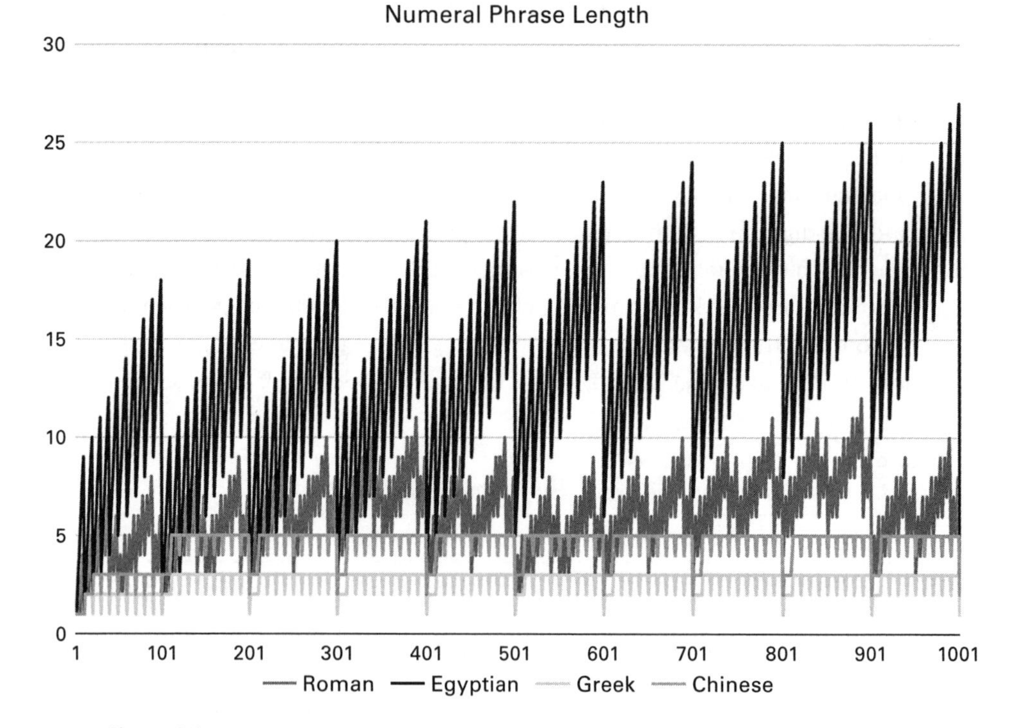

Figure 3.1
Size ordering of four numerical notations from 1 to 1,000

has mentioned it as a factor. But there are some theoretical reasons to imagine that it might be. One is that it might serve an approximative function—a well-size-ordered system gives a reader a rapid idea of the magnitude of a number, even when the number may have more digits than the four or so that subitizing can accommodate. Then, if two numerals have the same length (say, 2,479 and 6,127), we can immediately look to the leftmost digit to know which is larger. Similarly, in chapter 2, I showed how the property of conspicuous computation allows writers to violate assumptions of compactness to create visual effects on readers. Many of these texts depend on the materialized conceptual assumption that a long number is a big (and thus impressive) number. In turn, these seem to draw on deep-seated conceptual metaphors that underpin much of mathematics (Lakoff and Núñez 2000). The very fact that we describe numbers as "big" or "small" reflects the metaphorical structure of English and most human languages, whereby size is employed as a metaphor for quantity. This makes a lot of sense as a widespread metaphor, because three apples take up more space than two apples everywhere in the world. Finally, one of the widespread functions of numerical notation is the creation of ordinal lists, as part of organizational or administrative apparatuses. This is one of the features that Jack Goody (1977) identifies as likely to have cognitive effects in his discussion of the role of literate technologies in structuring knowledge. We might hypothesize that a size-ordered numerical notation more readily affords ordinal listing than one that is poorly size-ordered. But then again, one can have an ordered list that uses any notation whatsoever, and one of the few utilitarian functions for which we still use Roman numerals is creating ordinal lists—for instance, page numbers in the front matter of books, or subheadings in nested lists.

What we start to see, then, is that considering the properties of number systems in the abstract, without reference to what their users are actually doing with them, can be very misleading. One can probably always find a factor for which a system is very good or bad and then explain its success or failure *post facto* as a result of that factor. We need to go further and examine systems in use.

Screws, hammers, and Roman numerals

Let's imagine that you have a toolbox in your garage, full of all sorts of different useful things, and I'm your annoying neighbor. One day I drop by while you're working. I rummage around, pick up a screwdriver out of your toolbox, and say to you, "Gosh, that's not a very good hammer, is it?" Naturally, you protest that it isn't a hammer at all. Next, I hold the screwdriver by the head instead of the handle and say, "Well, of course, but you *could* use it like *this* to bang in nails. But that would be very cumbersome." You quirk your head, look at me, wondering whether I didn't hear you properly, and say to me, "No, really. It's not a hammer. I have a hammer, but it's in the trunk of my car, and that's not it." I respond obstinately, "Well, I've never seen your hammer, and it would really be a lot easier if you just used the handle of a screwdriver to bang in nails. Except that it's no good for that."

Now let's turn from this Pythonesque world to another scenario. You're an epigrapher and you find some inscriptions with some Roman numerals. You look at them and say, "Gosh, those things aren't very good for math, are they?" Of course, the writer is dead, so he/she doesn't say anything. Next, you fiddle around with the numerals and think to yourself, "Well, look at that! You *could* use those for arithmetic if you wanted to, but it would be very cumbersome." Again, the writer is not around to protest, although there are a bunch of texts describing the use of the abacus. You think of that, though, and say, "Well, it would really be a lot easier if they had just used numerals to do arithmetic instead of the abacus. Except that these Roman numerals are no good for that."

This second formulation has the very same structure as the first, and it is one that is quite widespread in the literature on the Roman numerals. In an important article, the psychologist Jiajie Zhang and the cognitive scientist Don Norman outline an argument just like the one above, as part of their claim that the Western numerals are universal today because they are better for arithmetic than other systems. They write:

> A more interesting case is the Roman counting board, in which counters were used for calculation and Roman numerals were used for representation, both in the same physical device. We propose that what makes the Arabic system so special and widely accepted is that it integrates representation and calculation into a single system, in addition to its other nice features of efficient information encoding, compactness, extendibility, spatial representation, small base, effectiveness of calculation and, especially important, ease of writing. (Zhang and Norman 1995: 293)

In other words, all other things being equal, they presume it is better to have a single system both for writing with numbers and calculating with numbers than to have specialized separate systems for each. But is it truly better?

I do not accept Zhang and Norman's premise that written numerals ought inevitably or automatically to serve a computational function. Rather, computation is potentially a function of numerical notation (to use Gibson's language of *affordances* that I discussed in chapter 1, numerals afford being manipulated for computation), but it is not an inevitable or natural one. I think that Norman, who is one of the central scholars in user-centered design, should properly be interested in how users of numerical notations actually employ them (Norman 1988). It is improper to judge a notation, or any other technology, in terms of purposes for which it was neither intended nor used.

In fact, among the thousands and thousands of Roman and post-Roman medieval Latin texts that survive, on a wide variety of media, none line up Roman numerals in columns for arithmetic, or use Roman numerals directly for arithmetical purposes. Admittedly, much of the material and literate record of of the Romans is lost, but there is no evidence to support the idea that reckoners in the Roman tradition ever did what we might imagine them to have done, which is to use Roman numerals as computational aids by writing them down and manipulating them. This has not stopped Western scholars from trying to show that it could have been done. I know of no fewer than five attempts, largely independent of one another, to show how the Roman numerals could have been manipulated to perform basic arithmetic adequately (Anderson 1956; Krenkel 1969; Kennedy 1981; Detlefsen et al. 1976; Schlimm and Neth 2008). All of these are artificial—they use the structure of the system and then imagine different ways it might have worked. None of them directly present any evidence to show that they were used in this way, because there simply is no such evidence.

At best, even when the conclusion is not so bad for the Roman numerals, it's ahistorical and strange to insist on these comparisons. We are so attached to the idea that numerals are for arithmetic that it's very hard to stop and ask whether they were actually used for doing calculations in a given society. Detlefsen et al. go even further; noting the absence of direct evidence for Roman numeral arithmetic, they argue, "If so, the existence of our procedures shows that the fault lay in the persons, not in their numeral system" (1976: 147). At worst, the unstated assumption of this framework is that the Romans were just bumbling along with inferior math for centuries, and then medieval folks just carried it on, either not realizing there was an alternative or not picking it up for some reason. For almost all numerical notation systems used over the past 5,000 years, there's precious little evidence that numerals were manipulated arithmetically in the way that Western people do. People often made use of a multiplication table written with numerals, or wrote the results of computations using numerals, but they didn't, by and large, line up numbers, break long numerals into powers to work with them, or anything of the sort. And since most people don't know that much about abaci and other arithmetic technologies, even though these technologies *were* obviously used for arithmetic, we assume (wrongly) that they certainly could never be as good as written numbers. And thus we conclude (wrongly again) that Romans were hopeless at arithmetic. We might even blame their (purported) lack of mathematical proficiency on their lack of a "good," "efficient" numeral system. So where does the truth lie?

For the republican and imperial Roman periods, most of our evidence for how arithmetic was conducted is inferential rather than direct. Of the thousands upon thousands of Roman inscriptions and papyri, we don't have texts where people wrote down Roman numerals in handy columns and worked on them directly. There are hundreds of tantalizing graffiti at Pompeii with Roman numerals, or with marks that use the I, V, and X of the Roman numerals strung one after another in long series, but none with what we can describe as an arithmetical calculation (Bailey 2013; Benefiel 2010). There are plenty of examples of individual numbers written on walls, but they often seem playful, or part of some numerical game, rather than an account of anything. This should not surprise us too much—after all, we don't normally do our arithmetic on walls, either. But the fact that there are so many written numerals on walls of houses and shops, and yet so little that looks like it is directly for computation, should give us pause.

In the absence of direct (epigraphic or paleographic) evidence showing us how the Romans did arithmetic, a very interesting study by Maher and Makowski (2001) surveys the literary evidence for antiquity, compiling a list of almost every classical Latin reference pertaining to computation, showing that, far from being backward in comparison to the Greeks, the Romans were adept at arithmetic and took great pride in that fact. They go further and suggest that, particularly when working with fractions, the Romans were fully capable of doing fairly sophisticated arithmetic that the authors believe would not have been readily possible using the abacus, which was decimal while the Roman fractions were duodecimal (base-12). It may strike the reader as preposterous that a numerical notation would have one base for whole numbers and a different base for fractions, but this was not especially strange in antiquity. The basic Roman fractions are cumulative-additive, base 12 with a subbase of 6, from 1/12 up to 11/12, and then with special signs for smaller fractions like 1/24, 1/48, all the way down to 1/288 (Maher and Makowski 2001: 397). Nor, they argue, was it likely that the widespread practice of reckoning on the fingers was used for fractions, because what we know of Roman finger computation has no evidence for the fractions (Williams and Williams 1995). Nor do the literary sources tell us anything about the modality through which fractions were manipulated in the Roman world. The Romans were clearly very adept at arithmetic, but no text, anywhere, shows evidence for working with Roman numerals to do so. Maher and Makowski's conclusion, then, is that the only means left, despite the lack of evidence, was the use of the signs for the fractions directly, for instance on a wax tablet. This is not ridiculous, but neither does it have any positive evidence in its favor.

Instead, the Romans, like the ancient Greeks, appear to have had what Reviel Netz calls a "counter culture"—a set of mathematical practices in which arithmetic is not written, but enacted through material culture such as beads on grooves (the classical abacus), tokens and counters, and, in financial transactions, through coins (Netz 2002). Netz shows that in the ancient Greek world, whose technologies are very similar and related to the Romans', the board abacus, coins, and *psephoi* (pebbles used in voting) were a cluster of technologies for manipulating quantities, whereas the Greek numerical systems (cumulative-additive acrophonic and ciphered-additive alphabetic numerals) were used principally for writing down numerical results. While we in Western contexts have fused these two functions, we run the risk of ethnocentrism if we presume that this is a natural or automatic way to use numbers. Unfortunately, much of the work on representational effects of numeral systems suffers from this flaw. Barring new evidence, we should presume that in Roman mathematics, numerical representation and calculation were relatively independent. Because Western arithmetic, beyond the level taught to preschoolers, is deeply dematerialized, it's all too easy for us to dismiss the materiality of tokens, abaci, and the fingers as "concrete" or inferior. Yet this universalizing and progressivist assumption is belied by the historical evidence that in most situations, numerical recording and numerical manipulation were conceptualized as complementary tasks for which different technologies could be relevant.

Even societies with lots of evidence for written arithmetic with numerals frequently used this as an adjunct to more material and artefactual practices like abaci. Karenleigh Overmann (2015, 2016), for instance, working on the numerical systems and arithmetical technologies of Mesopotamia, shows that a materially engaged arithmetic is nonetheless abstract and highly utilitarian. Christopher Woods (2017) further adds useful evidence that Mesopotamian societies, such as the Sumerians and Babylonians, used a kind of abacus—despite the lack of survival of the device itself, the textual evidence is substantial. Sometimes we know this from the kinds of arithmetical errors made in the texts, which reflect the structure of an abacus (a missing or extra pebble or bead) better than a written sign being transposed. In fact, most premodern societies likely had somewhat decoupled systems for arithmetic and numerical representation.

Nevertheless, the separation between notation and material technology was not complete. For instance, Roman abaci (figure 3.2) were usually inscribed with Roman numerals to indicate the values of columns as high as

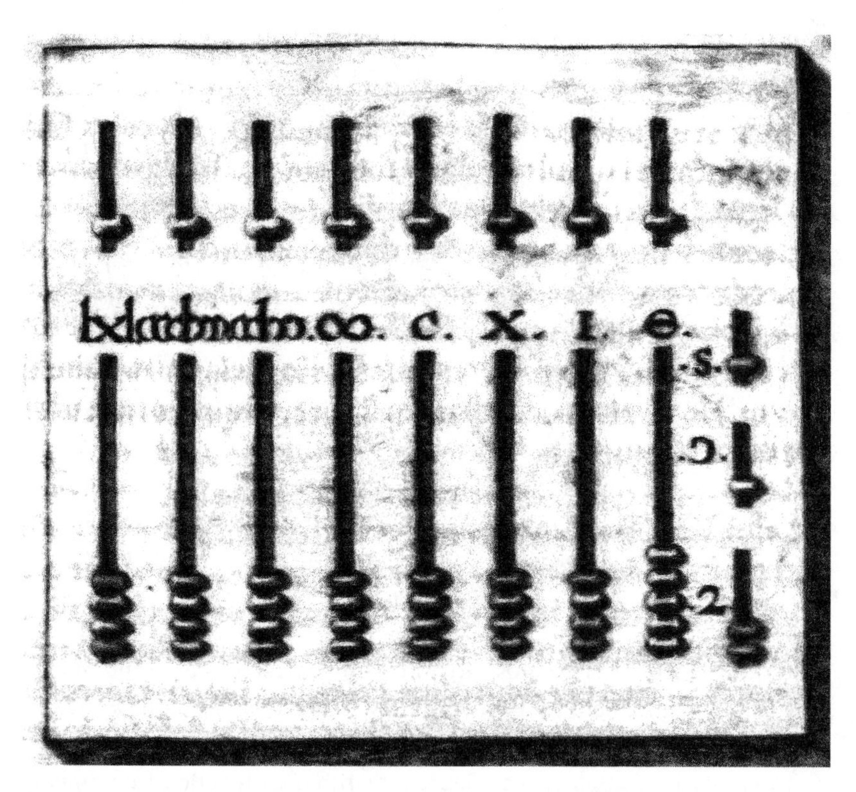

Figure 3.2
Depiction of abacus with Roman numerals (Velserus 1682: 819)

1,000,000 and including several fractional columns. And every abacus reckoner had to be able to translate the results both into Latin numeral words and into Roman numerals, as part of ordinary numerate practice. Taisbak (1965) even argues that Roman numerals originate from the abacus—in other words, that early Romans (or really, Etruscans and other peoples of Italy) started with the abacus and worked forward to a written number system. That's probably going too far, though. As Paul Keyser (1988) has shown, Roman numerals and Etruscan numerals actually originated from an earlier practice of tally marks. This is why the Roman numerals for the halves of the powers are visually "half" of the higher sign—V is the top half of X, L is the bottom half of C, and D is the right half of the original ↀ for M. They thus possess the property of relational, system-internal iconicity that Andréas Stauder (2018) describes for the Egyptian hieroglyphs—in other words, that signs resemble and point to other signs, more than to their referents in the world. While on their own V, L, and D are not iconic, through their visual "halving" relationship to X, C, and M, the whole system acquires a patterned set of meanings. They were not originally letters at all—otherwise we might expect X to be D for *decem*—and only became assimilated to the letters much later. These older, nonalphabetic forms for 500 and 1,000 persisted well into the modern era—we can still find them in the eighteenth century in the front matter of some books to indicate publication dates. As late as 1900, a monumental inscription was placed on the interior of the Class of 1875 gate at Harvard University reading ANN. DOM. CIↃDCCCC. COLL. HARV. CC↓XIIII, where we might expect MDCCCC (or MCM) for 1900 and CCLXIV for 264, with 1900 being the 264th year of Harvard's existence (McPharlin 1942: 18). In this very late instance, the use of the original, tallying/halving forms of the numerals 1,000 and 50 is probably purely an archaism introduced to add yet further prestige—anyone can write an ordinary Roman numeral, but this requires yet further cognitive work to be read, and indexes a past long forgotten, even at Harvard.

There is a material, artefactual form relevant to the development of the Roman numerals, but it is the tally mark, with notches or grooves repeated as necessary, not the abacus itself. Roman numerals are sometimes described as "tally-like," but that is not quite right. One of the things that differentiates a tally from a numeral phrase is that a tally represents an ongoing, sequential count of units. So, for instance, take the Roman numeral 28, or XXVIII. The repetition of X and I makes it look like a tally, but it's not, because one can't simply add more I marks on the end of it. A lot of the Roman tally marks described by Benefiel (2010) and Bailey (2013) scratched on walls at Pompeii look something like IIIIIIIIII or XXXXXXX. These are closer to tallies, but we would never mistake them for well-formed Roman numerals. They are more useful for one purpose, because they can be added onto as needed. But how those signs were manipulated and used, and how their use relates to the abacus, to the finger numerals, or to other, less well understood systems for manipulating number in the Roman world, remain mysteries.

So far I have discussed a number of Roman and post-Roman medieval arithmetical techniques: finger arithmetic, the abacus, tallying. To this we could add mental and verbal arithmetic, in whatever form they took. The study by Maher and Makowski (2001) discussed above, which summarizes essentially all the major references to arithmetic from classical Roman texts, shows just how nebulous is the effort to directly attack the problem. We are hampered by a woeful lack of material evidence; for instance, there are only four physically surviving hand-held Roman abaci (Schärlig 2004). Now, there is no doubt that there was an abacus in Roman arithmetic. Numerous texts discuss its use. There are artistic depictions of reckoners using the device, although the representations themselves are rather vague. Other texts have numerical errors best explained through a misplaced counter—i.e., the quinary-decimal structure of the board explains the error (Laroche 1977). But while we know it was used, much remains unclear, including the algorithms or processes by which calculators did their work.

We do know a little more about medieval arithmetic on the counting board because many medieval and early modern arithmetic texts discuss it. The medieval counting board is not a direct offshoot of Roman practice, and unlike the pebble-in-groove classical technology, the counting board or table was normally a board or table with lines, or a marked cloth covering placed on a hard surface. Loose *jetons* or counters (usually blank) were placed on the cloth and manipulated as needed. Figure 3.3 is a jeton (around 2.5 cm in diameter) from 1553 from Nuremberg, on which the act of reckoning on the board is depicted.

A recent study by Cheryl Periton (2015) applied a historically informed practical account to the medieval European counting table. Periton's study involved, effectively, teaching herself to work with the board, not based purely on introspection but on the large number of early modern arithmetic

Figure 3.3
Jeton from Nuremberg, 1553, a counter for use with the medieval reckoning board (Rekenpenning Neurenberg rekenaar 1553 by Kees38 is licensed under CC BY-SA 3.0; source: Wikimedia Commons)

texts that taught its use. She found that, like East Asian computing devices, the counting board was not especially difficult to learn at a basic level, although expertise would have taken many, many hours. One advantage of these sorts of mobile boards, where pebbles or counters are not fixed to a column, is that counters may be more readily shifted as needed from one place to another—there need not be any "unused" counters because the calculator places them only where they are needed. Periton does not show, and I do not think it is possible to show, that arithmetic on counters is faster, or cognitively superior, to written arithmetic or any other arithmetic. Because we have no Romans or medieval Europeans around to tell us otherwise, this leaves open the possibility that arithmetic on boards and beads is not really that useful. But all is not lost, because the Greco-Roman abacus, though apparently historically unrelated to the East Asian abacus, has much the same structure. And in East Asia, we have ample evidence for a flourishing, rapid and accurate materially engaged computation system.

Throughout the twentieth century, East Asian abaci have held a certain fascination for Western audiences—especially Americans. A popular story is recounted in many histories of mathematics. On November 12, 1946, at the height of the American occupation of Japan at the end of World War II, a competition was held in Tokyo between Kiyoshi Matsuzaki, an expert user of the *soroban* (the Japanese variant of the East Asian abacus) and an American, Private Thomas Nathan Wood, trained in the use of what was (for the time) a state-of-the-art calculating machine, across the four basic arithmetical functions, on the basis of speed and accuracy (Adler 1954; Kojima 1954). The *soroban* user won the competition handily, much to the dismay of *Stars and Stripes* magazine which had sponsored the event, presumably with the goal of demonstrating the superiority of Western technology. But this competition was not an isolated incident—it was one of several such contests held by Americans between abacus users and users of adding machines, the earliest of which goes back to 1910:

> At a meeting of the Tacoma Chapter of the American Institute of Bank Clerks the other night a Japanese clerk, using a "saroban," "put it all over" a Tacoma bank clerk using a modern adding machine in casting up a long column of figures. The Japanese, says the Tacoma Ledger, with his "saroban," did the arithmetical "stunt" in thirty seconds, while the bank clerk, with his modern adding machine, took fifty-four seconds to obtain a total, and afterward it was found the clerk with the modern adding machine had made an error, while the Japanese with his little abacus or "saroban" obtained the correct result the first time. (Anonymous 1910)

While even the best electronic calculators of that time may not have been much faster than pen and paper, the notion that the counting board is somehow inferior must be discarded. What these sorts of examples show us is that Americans, at least, felt a powerful urge to compare the abacus to modern computation. It's the same sort of obsession that leads us to think that these sorts of comparisons are actually the basis on which we choose a numerical notation system.

There is ample psychological evidence that users of counting devices like the Chinese *suan pan* and Japanese *soroban* (both of which we somewhat errantly and ethnocentrically label "abacus" in most Western languages) do not suffer at all from this separation of computation and representation.[4] In the 1980s, a series of papers by both Japanese and American scholars investigated the arithmetical practices of East Asian experts in the *soroban*, showing that the arithmetical capacities of contemporary East Asian abacus experts are equal to or surpass the arithmetical abilities of users of pen-and-paper algorithms (Stigler 1984; Hatano and Osawa 1983). Abacus masters are able to construct and use a "mental abacus" in the absence of the actual device, just as people who are well trained in pen and paper arithmetic can visualize and mentally work with Western numerals.[5] We thus ought to be skeptical that an integrated system is, in practice, the best one.

Recognizing that one cannot predict arithmetical competency from technology alone, the East Asian evidence rebuts the idea that an integrated computational-representational system should presumptively be considered superior to a division of cognitive labor. It is as if we were to insist that, because a screwdriver can be used to drive in nails by turning it around and using the handle as a makeshift hammer, we have no need of hammers. The evidence above should give us pause, but our cognitive bias in favor of the functional association with arithmetic can overpower it. This doesn't absolve us of an obligation to be aware of the cognitive properties of systems. The properties of conciseness, sign count, and extent all are relevant not only for computational functions but also for representational ones. Representation is a universal and defining property of numerical notation, because it is impossible to imagine a written numeral system that does not represent numbers that are intended to be read. It thus serves as a better place to start when looking for general factors explaining the transformation and replacement of numerical notations. But if we start with a bias toward computational functions, the way we look at and evaluate those factors will be skewed toward properties of notations most relevant to arithmetic.

This is a casual, all-too-easy ethnocentrism, and hard to detect. It's not the nativistic, "our ways are good, your ways are bad" ethnocentrism that we hope we mostly know to avoid in the humanities and social sciences. Because arithmetic as it is presently taught almost everywhere relies on the structure of the positional decimal numerals, lined up and manipulated as needed, this practice takes on a naturalness that is deceptively difficult to untangle. Yes, the Roman numerals are relatively difficult to use if you presume that the way to use them is to break them apart, line them up, and do arithmetic in something like the way we were taught. This isn't to say that the functions of technologies aren't relevant, but if we decide in advance what their functions must be, we are likely to miss what they actually were. To hammer the point home: if we do that, we're screwed.

Awareness and metanotational commentary

When linguists and anthropologists investigate the intersection of language and culture, they frequently address questions relating to *metalanguage*—cultural beliefs, discourse, and behavior that refers to or concerns language itself. Also known as *language ideology*, metalanguage comprises local cultural models of language that may be expressed, contested, or taken for granted in particular contexts.[6] Whether they are minor—such as whether the English word donut should be spelled *d-o-n-u-t* or *d-o-u-g-h-n-u-t*—or major—such as ideas about the value of minority and politically subordinate languages—metalinguistic beliefs shape how people understand the linguistic aspects of the world around them. In turn, these evaluations shape how people make decisions about what to say (and when and where), and what speech to find acceptable or unacceptable. Analyzing metalanguage can thus be a valuable tool for understanding how languages change.

Similarly, with regard to numeral systems, we might ask how their relative efficiency is perceived and talked about by users. Gareth Roberts (2000) discusses how, in the nineteenth century, the evaluation of the comparative merits and defects of English and Welsh verbal numerals became a political issue, as shown in the following material from the nineteenth-century broadsheet paper *Seren Gomer:*

> *Dichony Sais rifo mil yn araf deg cyn y gallaiy Cymro, er poethi ei geg gan frys, rifo pedwar cant.*
>
> An Englishman may count to a thousand slowly before the Welshman, although his mouth may overheat in the attempt, can count to four hundred.
>
> *... rhaid i'r Cymro, oni fydd yn go gyfarwydd, fyned i fyfyrio cryn dipyn cyn adrodd ei rif; dichon ddywedyd y nod cyntaf, sef tri chant, gystal a'r Sais, eithr pa le y dichon y peth gwirion gael dau ar bymtheg a thrugain?*
>
> ... the Welshman, unless he is particularly expert, has to think long and hard before reciting his number; he may be able to recite the first digit, three hundred, as well as the Englishman, but how on earth can a foolish man achieve two on fifteen and three twenties?

Welsh today is characterized by two numeral systems, the older of which has various irregularities and has many base-20 components, and the newer of which is purely decimal and has no irregularities. So, for instance, eighty-seven in the traditional system would be *pedwar ugain a saith* (almost literally "four score and seven"), whereas in the new system it would be *wyth deg saith* (eight ten seven). Roberts convincingly argues that the negative comparison of the older system with English was one of the factors promoting the newer Welsh system in the nineteenth century. In the specific context of the *Seren Gomer* article, the debate was contested in, of all things, a counting contest, in which proponents of the two number systems started to count up from "one" in their preferred lexicon, to see who could count fastest. So the narrative goes, the pure decimal counter won the day, demonstrating its greater efficiency. We need not take this story at face value, and it certainly did not settle the issue—both systems continue to be learned and used regularly in Wales even today. What it does show is that very often people do care very much about their number systems, and use metalinguistic commentary to reflect on these sorts of issues.

We might similarly ask how the users of numerical notations perceived the systems they were using, in order to assess the degree to which they were aware of and considered relevant particular features or defects. Commentaries and critiques with a metalinguistic (or in this case, metanotational) purpose allow us to get beyond speculation about which factors were relevant to the replacement or abandonment of a system, by using contemporary texts to establish the relevance of particular criteria to their users.

Unfortunately, very few premodern texts discuss the cognitive advantages of a new system, or the defects of existing ones. Where they do, often the new numerals are praised in extremely generic terms. The Syrian scholar Severus Sebokht, who was one of the first to describe ciphered-positional Indian decimal numerals systematically in 662, wrote of the Indians' "subtle discoveries in the science of astronomy, which are more ingenious than those even of the Greeks and Babylonians, and their method of calculation which is beyond description—I mean that which is done with nine symbols" (Burnett 2006: 15). This is high praise, but "beyond description" indeed—so much so that there is no description to help us! But conversely, centuries later, the Arabic poet and chronicler Abu Bakr bin Yahya al-Suli (880–946) indicated a metanotational preference for finger computation over working with numerals for arithmetic, because of the ability to keep results secret, which preserved the value and dignity of the scribe (Kunitzsch 2003). The mathematician al-Uqlidisi (ca. 920–980) appreciated Indian numeration, noting that "it is easy and quick and needs little precaution," because with written arithmetic one is less susceptible than with finger reckoning to losing track of one's computations; at the same time he noted that this new technique, *hisāb al-hindī*, was much disliked and so the writer should perform these calculations on the dust board (*takht*)—a flat board covered in sand on which one could write—and then when writing the result on paper use the Greek or Arabic (ciphered-additive) alphabetic numerals (Saidan 1966). This is not the argument of someone enamored of this novelty, but rather someone who sees multiple techniques and notations having different uses in different media.

In the Indian subcontinent there has always been a tension between the longstanding written tradition and a strong preference for oral learning; numerals are often written out in full in Sanskrit verse because these were meant to be read rather than simply manipulated. Babu (2007), investigating mathematics education in eighteenth- and nineteenth-century *tinnai* "veranda schools" in Tamil-speaking parts of South India, emphasizes the continuing role of oral transmission of mathematical information and the emphasis on memorization and mental arithmetic. This is a similar argument to that raised over a thousand years earlier by Socrates, among others in the classical world, in favor of oral knowledge and memorization against literacy (Small 1997). Such comparisons are from traditions in which the older numerical systems were ciphered-additive and alphabetic, with a different letter assigned to each multiple of each power of 10, whereas the newer notations were ciphered-positional and not linked to local scripts. By the tenth and eleventh centuries, both of these traditions, as well as finger reckoning, as well as mental arithmetic, had coexisted for hundreds of years in India and the Middle East. The continued preference for finger reckoning should tell us something immediately about the way that positional decimal numeration was perceived relative to its competitors. Strikingly, with one possible exception there are no surviving Arabic texts prior to the eleventh century that use positional numerals at all (Burnett 2006). There are literary references that show us that they must have been used, but the actual use of signs for 0 through 9 is not attested until almost four centuries after they had been first introduced from India. Just as the Roman numerals took centuries to be replaced, the older Arabic traditions took centuries to be replaced and, like Roman numerals, continue to be used by some to the present day.

Even more striking than the absence of positive endorsements of specific features of ciphered-positional decimal numeration is the absence of negative evaluations of the defects of existing systems. Durham's (1992: 40–41) discussion of medieval European accounting practices notes that "one does not hear complaints from Medieval merchants or bankers concerning the quality or speed of the arithmetic available to them, nor is there dissatisfaction with the Roman notation." In part, this is because the Roman numerals were not used for arithmetic in the same way that Western numerals can be—as discussed above, they are a representation system, but not a computational technology.

In Western Europe, starting around the eleventh century, there were two basic techniques used for arithmetic. The first, employing the abacus, extended the reckoner's cognitive work using a flat board and counters as described above. The second was the technique of *algorism*, the use of the Western numerals 0 through 9 using place value—not quite as we do today, but still a form of written arithmetic using ciphered-positional numerals. Both techniques had advocates and detractors, and this has been characterized, somewhat unfairly, as a centuries-long debate between "abacists" on the one hand and "algorists" or "algorismists" on the other. To be sure, there were proponents of each position, with texts emphasizing one over the other. But few if any of these texts cast aspersions on the other method. Leonardo of Pisa (Fibonacci), whose 1202 *Liber abaci* would suggest by its name that he was an abacist, was in fact perhaps the most vociferous advocate for the algorism (Burnett 2006). True, after Fibonacci, *abacus* texts placed heavy emphasis on computation with Western numerals (Stedall 2001). Reckoning with algorism did become more common among the learned, but this was not at the expense of reckoning on the counting board, which continued to be taught and advocated well into the seventeenth century both in commercial mathematics and in material of a more scholarly bent. In a recent article, Nothaft (2020) argues persuasively that casting this as a debate between competing factions is a mischaracterization, and that modern historians of mathematics have overemphasized this contestation. Even more pernicious, he argues, is the occasional claim that users of the abacus and Roman numerals treated the zero and the Western numerals as heretical or diabolical, a claim for which there is no evidence whatsoever.

In any event, whatever debate there was was not really a debate about numerical notation at all. Both abacists and algorists continued to use the Roman numerals widely for all sorts of functions in texts. In fact, the jetons that the abacists used could be marked with either Roman or Western numerals, and for some writers, such as Pandulf of Capua, an eleventh-century Italian abacist, the default assumption was that the abacus was to be used with the Western signs, excepting the zero (Gibson and Newton 1995: 308–310). While the use of abaci (or counting boards in general) was the subject of discussion, abandoning the Roman numerals was not seen as part of this discussion. This should immediately tell us that whatever the debate was about, it wasn't really about numbers but about techniques and technologies.

When the Western numerals were first being introduced into Western Europe—transmitted through Italy and Spain, both of which were in extensive contact with the Muslim world—they were not seen as a general-purpose notation system. Rather, their adoption was discussed particularly in the context of arithmetic—as a choice between pen-and-paper arithmetic with Western numerals or computation with tokens on the medieval counting board. For everyday purposes, many of the most notable mathematicians used Roman numerals throughout their writing, without any complaint or concern. And why should they have cared? After all, Euclid and Archimedes and all the other Greek mathematicians of antiquity achieved all that they did without the benefit of place value.

The first source I know of to specifically compare the Roman and Western numerals and find the former wanting is the 1275 *Doctrina pueril* by the Catalan theologian and logician Ramon Llull (ca. 1232–1315). Llull is today known for his contributions to mathematics, specifically to theories relating to voting and elections. The *Doctrina pueril,* as its name suggests, was a text for the education of young people, and among its many topics is a discussion of arithmetical practices (Dagenais 2019).[7] Chapter 74 contains a remarkable passage on the use of numbers:

> This art exists so that men can better know how to hold a number in memory and in physical sight, for the nature of memory is that it forgets many things more readily than a single thing. For that reason, written numbers were created, that is: X, XX, XXX, C, M, MM, CM. And when these figures are not sufficient for writing what you need to, you can turn to ciphers and to algorithms using Arabic numerals or to the abacus, which are more easily read and understood.

Llull is among that generation of mathematically proficient scholars who, in the wake of the publication of Leonardo of Pisa's *Liber abaci* in 1202, were becoming increasingly aware of and familiar with ciphered-positional decimal numerals, written rather than manipulated on tokens on boards, as a normal practice. Yet Llull does not see the Roman and Western numerals in conflict;

rather, he is clear, first, that the Roman numerals (which would not have been called Roman at the time—they were so universal that they were simply "figures") are the default, ordinary choice, and that they do serve important mnemonic purposes. Llull notes, though, that they are not "sufficient" in all cases and then recommends, as a secondary option, the Western numerals, "which are more easily read and understood."[8] So this is clearly an instance of metanotational commentary in which the Roman numerals are recommended, but then at the next moment criticized. In writing this document for the education of the young, Llull does not describe or list the Western numeral signs or discuss their use further—merely noting their existence. It is certainly not an argument for the replacement of the Roman numerals. Nor does it focus on arithmetic at all—rather, his account concerns memory, clarity, and readability, factors that could be relevant for arithmetic, but which are also relevant for every other function for which numerals could be used.

Next, consider the allegorical woodcut *Typus arithmeticae* of Gregor Reisch, published in 1503 in his treatise *Margarita philosophica* (figure 3.4). This figure, reproduced in dozens of histories of mathematics, is often taken as an icon of the debate between notations. As Shana Worthen (2006) shows, analyzing how medieval iconography represents and conceptualizes inventions can be extraordinarily insightful in understanding how these innovations were understood. Here, we see the allegorical figure of Arithmetic, her garments bedecked with Western numerals, looking upon two reckoners. On the left, we have the esteemed mathematician of late antiquity, the sixth-century Roman Boethius, using numerals to compute. (Ironically enough, the historical Boethius could not possibly have known Western numerals, which were centuries away from being known in the West at the time!) On the right sits Pythagoras, using a Greek-style abacus with pebbles and grooves. The numerals on Arithmetic's garments, and her head turning in the direction of Boethius, signal her approval. But note what is missing. There is not a single Roman numeral to be seen anywhere in the woodcut. In 1503, when Reisch made the *Typus arithmeticae,* Western Europe was already rapidly beginning to replace the Roman numerals with the Western numerals—this was the height of the period where we might expect them to be compared directly. But the comparison here is about forms of computation, rather than forms of numeration. Because the Roman numerals were not really used for arithmetic in this way, including them would have missed the point.

Figure 3.4
Typus arithmeticae, an allegorical woodcut by Gregor Reisch (1503)

In the late 1680s, the English folklorist and antiquary John Aubrey (1626–1697) reflected on the decline of the Roman numerals, both regretting their loss while aware that they are a product of a bygone era. Aubrey's *Remains of Gentilisme and Judaisme,* a set of short antiquarian reflections, contains the following passage:

> All old accounts are in numerall letters. Even to my remembrance, when I was a youth, Gentleman's Bayliffs in the Country used no other, e.g. i, ii, iii, iiii, v, vi, vii, viii, ix, x, xi, etc. and to this day in the accounts of the Exchequer. And the Shopkeepers in my Grandfathers times used to reckon with Counter: which is the best and surest way: and it is still used by the French. (Aubrey 1881: 124)

Here, by "numerall letters" Aubrey means the Roman numerals—this was a fairly typical appellation at the time, as they were not particularly seen as Roman in seventeenth-century England.[9] Writing in the last quarter of the seventeenth century about events fifty years earlier, he was witness to a period of rapid transition. He notes, as I have already argued, the division of labor between numerical representation (using Roman numerals) and computation (using counters / the counting board), both of which were widespread throughout the sixteenth and seventeenth centuries but which were giving way to a more modern arithmetic with Western numerals on paper. But rather than seeing this as an unmitigated good, Aubrey insists on the superiority of the older system of reckoning, and seems to lament the gradual loss of older notations.

Around the same time, however, Aubrey's colleague, the mathematician John Wallis (1616–1703)—both authors were founding members of the Royal Society—offered a different perspective. Wallis's *A Treatise of Algebra Both Historical and Practical* (1685) is a historically informed survey of English mathematics that pays particular attention to the Western numerals as transmitted from the stock of Arabic learning (Stedall 2001). Wallis writes:

> Before these Figures were introduced, while we had no other ways of Notation for Numbers than that of the Latin, by a few Numeral Letters, M D C L X V I; or of the Greeks by the Letters of the Alphabet, α, β, γ, δ &c. (like as before them, the Hebrews, Arabs, and other Orientals, did also design Numbers by the Letters of their Alphabet). The exercise of Practical Arithmetic, especially in large Numbers, was but very lame, in comparison of what now it is. (Wallis 1685; Stedall 2001: 88)

Wallis makes it clear that he regards the transition away from additive "numeral letters" like the Roman numerals to have been a great advance for mathematical learning. We know quite a bit about Wallis's experience with numerals through his autobiography, where he describes encountering the Western numerals as a boy of fifteen when one of his younger brothers (destined for a trade) had been learning the new arithmetic and numerals (Otis 2017: 453). Unlike Aubrey, Wallis's attitude embodies a modernizing, reformist attitude that links arithmetical reform and numerical reform together.

Finally, in 1701, the Scottish doctor and polymath John Arbuthnot published his *Essay on the Usefulness of Mathematical Learning*, remarking:

> The nations that want it are altogether barbarous, as some Americans, who can hardly reckon above twenty. And I believe it would go near to ruin the Trade of the Nation, were the easy practice of Arithmetick abolished: for example, were the Merchants and Tradesmen oblig'd to make use of no other than the Roman way of notation by letters, instead of our present. And if we should feel the want of our arithmetic in the easiest calculations, how much more in those that are something harder; as simple interest and compound, annuities, &c., in which it is incredible how much the ordinary rules and tables affect the dispatch of business. (Arbuthnot 1701: 19)

For Arbuthnot, this is not an issue of memory or of cognitive efficiency—rather, numerals are the product of, and themselves produce, civilization through their commercial utility. His is a progressivist perspective that explicitly contrasts English arithmetic with that of Native Americans, whose purported limited numerical lexicon and lack of facility with numbers was quickly becoming a modernist trope. In this account, the Roman numerals were perhaps an intermediate step on the path to civilization, but nothing more—best abandoned. As Michael Barany (2014) has shown, this "savage number" orientation of European scholarship was used to justify pseudo-evolutionary and frankly racist notions about the cognitive capacities of non-Western people. This perspective would reach its peak in the heady days after the publication of Darwin's *The Origin of Species* (1859), as thinkers began to apply newly developing evolutionary ideas to the growing ethnographic record at the time, with John Crawfurd's 1863 essay "On the Numerals as Evidence of the Progress of Civilization" (Crawfurd 1863).

Wallis and Arbuthnot present us with clear discourse in which the Roman numerals are argued to be an inferior notation with a negative effect on arithmetical practices. But these are very late examples relative to the history of the numerals themselves—the Western numerals had coexisted in Europe for five hundred years by this point without any significant complaint. And, as these authors themselves note, the Roman numerals had been largely replaced throughout Western Europe by this point. By 1700, while the Roman numerals were still in use, the eventual predominance of Western numerals was hardly in doubt. So we do not have grounds to argue for causation—the texts are far too late. Rather, this sort of discourse is a justification after the fact of what had already happened. Along with this progressivist perspective, they also embody another aspect of the rationalist elements of "Enlightenment" thought, with its emphasis on logic and functionality. Under this view, if something was replaced, it probably wasn't that good to begin with.

Yet even so, despite these early modern commentaries, the widespread description of negative features of Roman numerals is largely a twentieth-century phenomenon. Using Google's Ngram Viewer to track word frequencies over time in English books and magazines, I examined the frequency with which the term "Roman numerals" was preceded by one of the adjectives "cumbersome," "clumsy," or "awkward" (Chrisomalis 2017). Figure 3.5 shows the proportion of all examples of references to Roman numerals, in each year, in which one of these descriptors preceded it. In the nineteenth century, almost no one was using this framework for judging Roman numerals, but then, starting especially around 1900, it became much more common. Many of these references are in arithmetic textbooks, popular histories of mathematics, and scientific sources. Note, too, that after World War II this discourse declined—although it is still much more common

Figure 3.5
Proportion of instances where "Roman numerals" in English books is preceded by "awkward," "clumsy," or "cumbersome," 1800–2000 (Google Books Ngram Viewer, http://books.google.com/ngrams)

than it was in the nineteenth century. Perhaps by that point the Roman numerals had become rare enough in everyday life that they no longer attracted quite so much negative judgment.

Far from being an eternal perspective on the Roman numerals, metanotational discourse arguing that they were inefficient or cumbersome was rare in the nineteenth century, became much more common in the early to mid twentieth century, and then once again declined. A sizable proportion of these modern texts are, perhaps unsurprisingly, histories of mathematics. This sort of discourse provides a rationalization of a historical fact, projected onto the medieval mind, rather than an explanation of that fact. While that doesn't mean that issues of efficiency weren't relevant to the Roman numerals' decline, it should give us pause as we think about why it is that we might have believed it to be so in the first place.

From awareness to causation

Finally, we need to consider what kinds of evidence would show us that the awareness of the advantages or deficits of a particular system actually did produce the changes that were observed. Even if there is some awareness of some disadvantages of some technology, this does not immediately militate against its use. People hang onto technologies all the time, even when they are aware of alternatives that might be superior. To draw on a notational comparison, consider the various efforts to promote the conversion of Chinese script to Roman alphabetic letters throughout China, including many developed by Chinese scholars, such as pinyin (DeFrancis 1984). The discourse from Chinese speakers stressing the mnemonic challenges of the characters relative to a phonographic script is staggering. Yet Romanization has yet to break through significantly in East Asian education, in large part because of the cultural value attached to the use of characters. Similarly, the deficiencies of the QWERTY keyboard have often been discussed, but this has not led to its replacement by the Dvorak or other "superior" keyboards, in this case because of the learning and other costs involved in switching.

Because there is so little discourse about numeral systems in the historical record, we might be discouraged. However, there is a partial solution if we look at systems not through a synchronic lens but diachronically, attending to particular transformations in systems that suggest efforts to improve one or another feature of a system. People may not be telling us what they think, in other words, but the sorts of actions they take to change their numerical systems tell us much about problems they might perceive, and how the solutions fit in with their beliefs.

The Roman numerals, for instance, originally topped out at M, so that there was no ordinary way to express 5,000 or higher numbers. New signs for 5,000, 10,000, 50,000, and 100,000 were introduced around the late third century BCE, though they were rarely used (Keyser 1988). In the terms described above, this transformation addressed issues of extent. But then, in the late republican period (the second half of the first century BCE), the Romans began using the *vinculum*, a horizontal line placed atop a numeral or numerals to indicate that it should be multiplied by 1,000 (Gordon 1983). This wasn't used in addition to the signs for 5,000 and higher, but rather, replaced them, so that in place of ↂ for 10,000 a writer might write X̄ instead. This shows a concern with sign count—the rare and obscure signs for 5,000 and higher were effectively replaced with a single multiplicative sign that could combine with the ordinary Roman numerals I, V, X, etc. A century or so later, a new notation was introduced by which a numeral surrounded by vertical bars on either side and a horizontal line on top (i.e., a box open at the bottom) indicated multiplication by 100,000. Thus, the *columna rostrata* inscription I discussed in chapter 2, with endless signs for 100,000, could have been replaced with a short numeral phrase (probably XXXII = 32) with this multiplicative mark around it. In this text, doing so might have defeated the purpose of ideologically marking the victory with an immense and impressive number, but in plenty of other texts that wasn't a concern. This change shows a concern both with extent and conciseness.

Other transformations in the Roman numerals were rare, local, and used only by a handful of writers, as the need arose, disappearing almost as quickly as they were invented. One of these was the inventive positional Roman numerals developed by Ocreatus, a student of Adelard of Bath, in the early twelfth century (Burnett 1997: 43–45). Ocreatus had learned of the zero and place value and then simply combined them with the Roman numerals I, II, III ... IX, so that 1089 would be I.O.VIII.IX. No author seems to have taken Ocreatus up beyond this initial attempt. But that they were invented at all suggests that Ocreatus recognized that adding a zero sign allowed the expansion of Roman numerals to express any integer, producing a truly infinite system that combined cumulative-additive notation within each power of the base, with positional notation across different powers. Similar sorts of transformations were effected, with little long-term consequence, by authors experimenting with positional variants of the (normally ciphered-additive) Hebrew and Greek alphabetic numerals, by adding a sign for zero alongside the ordinary signs for 1–9 in those systems (Schub 1932). Similarly, under the influence of Arabic knowledge in Spain, various efforts were made to turn long strings of repeated numeral signs into one, so that 40 (XL) became ligatured together into a single sign, and in place of the ordinary Roman numerals for 3–9, some manuscripts used single characters, such as the first letter of the number words (the acrophonic principle) (Lemay 1977). These sorts of transformations suggest that for at least a few users, conciseness was perceived to be advantageous and the increased sign count was not a concern. However, these hybrid notations were never adopted more widely, so the concern cannot have been sufficiently widespread to create a lasting change.

These developments, which occurred among writers familiar with Western or Arabic place value numeration, give us a sense of the features of the new system that may have appealed to them, and thus provide insight into both how the existing (Roman) and new (Western) systems were perceived. These transformations are not patterned cross-culturally in a way that would allow broader generalizations, but they do give us relevant information about local contexts in which one or another cognitive factor was relevant to a notational change. They also show us the extent to which users could and sometimes would go to modify rather than replace an existing notation. The outright replacement of the Roman numerals was thus simply the most extreme consequence of whatever set of factors motivated medieval writers to think about notations in the first place.

It is vital that representational effects of numerical notations be considered in their historical and social contexts of use, rather than as pure abstract structures from which particular cognitive consequences flow. In this, I am echoing some of the best ethnographic and cross-cultural psychological work ongoing in modern social settings, of which Geoffrey Saxe's work with the Oksapmin of Papua New Guinea is our best contemporary example (Saxe 1981, 1982, 2012). Saxe, a developmental and educational psychologist, began his career looking at the structural features of the Oksapmin body-counting system—a numbering system which traditionally began counting "one" on one thumb and proceeded along one arm, up across the shoulders and head, and down the other, with various body parts sharing numerical referents. From the sorts of evolutionary and progressivist frameworks of many developmental psychologists, Saxe's early work is open to a reading as a sort of claim for concreteness of numerical cognition among the Oksapmin. But upon returning to the Oksapmin for new fieldwork and experiments after an absence of over twenty years, Saxe instead found himself compelled, not by what the Oksapmin system had been, but by how it had transformed in light of changes to economic, educational, and social systems. Rather than seeing the Oksapmin as prisoners of an inferior notation—which was never his claim—Saxe was able to show how the body-counting system shifted in both structure and contexts of use as global economic and social forces impinged on Oksapmin lifeways. Notations and techniques such as these are not static entities that index discrete mental stages; rather, they are tools and resources that accompany others (such as language) in facilitating different sorts of numerical cognition.

This tacking back and forth between the cognitive, the social, and the material echoes the important arguments of Jocelyn Penny Small (1997), who links memory as a cognitive process to the techniques of literacy in the Greco-Roman world. The material culture of literacy, in this view, cannot be treated on its own, without reference to the minds that produced and consumed texts. But neither can the psychological studies themselves be predictive or fully explanatory, because cognition is extended and expanded—we think with the material world, not just about it (Clark 2008; Menary 2010). But still this is not enough—because people do think about their tools, and reflect on their properties, and talk about them (metalanguage and metanotation). So, for instance, how Greeks and Romans thought about mnemotechnics (the technologies of memory) should greatly influence how we understand their texts (Small 1997: 82–97). For numerical systems, the implication is not that we should forget about numerical cognition, but that we must be attentive to the entire social-material-ideological system in which these properties became relevant.

Ethnographic research on numerical cognition benefits from having people at hand to talk to, to work alongside, and to offer cognitive tasks to. But once we step back from modern times, we require historical approaches and methods if we hope to make verifiable claims. The tools of the experimentalist and the ethnographer are not available to us. We don't know when the last communities of active Roman numeral users ended their practices. Bazzanella, Kezich, and Pisoni (2014) discuss the fascinating numerical practices of shepherds in the Fiemme Valley (Trentino, Italy), where Roman numeral red hematite graffiti on stone persisted into the twentieth century alongside Western numerals. But so, too, has the practice of Roman numeral tattoos (e.g., of a significant date) become newly popular in the West over the past decade. These inscriptional practices are interesting but don't reflect communities where the numerals are used regularly or typically—and certainly not in anything like an arithmetical context. Similarly, we can't assume that particular cognitive factors are always relevant—otherwise, why would the Roman numerals have survived as long as they did? But we also can't rely simply on metalinguistic or metanotational accounts, because there are so few of them. We also need to be aware of other sorts of factors that might be noncognitive, or not cognitive in quite the same way as described above, that might bear directly on the question of why users would switch from one system to another, or why, over decades or centuries, the Roman numerals would fall increasingly out of use in more and more contexts. To do that, we need to look at what actually happened in the key period in early modern Europe when the Roman numerals finally fell from favor.

4 / IV The decline and fall of the Roman numerals, II: Safety in numbers

Consider the following three fictional but familiar examples of decision-making processes:

A. So many golfers are using Whackalot clubs these days. They must know something I don't about the quality of those clubs. Anything that is so popular must be better than less popular alternatives, so I think I'll pick myself some up Whackalots.

B. Everyone is wearing zipper-pocket jeans this year. They are so cool! Why, just the other week, I saw ten people wearing them in my linguistics class. I want to fit in with the crowd and be cool, so I'm going to pick up some zipper-pocket jeans.

C. Huge numbers of people use Headshot as a social networking website. If I use some other website instead, I won't be able to chat with all my friends. Therefore, I'd better give in and start using Headshot.

In each of these cases, a decision-making process to adopt something is initiated or accelerated by the information that this thing is commonly used. The reasoning underlying each decision, however, differs considerably. In this chapter, I will analyze why a single characteristic—frequency of use—and a single outcome—the adoption of a novelty—should interact in these three radically different ways, and apply this to the particular case of the Roman numerals in late medieval and early modern Europe. As we saw in the previous chapter, explanations that are based on abstract consideration of the structure of the Roman versus the Western numerals aren't sufficient to explain why people chose to abandon one system and prefer a new one. Decisions are cognitively motivated, but not in ways that can be predicted simply. Here we'll look at what the actual process of the replacement of the Roman numerals looked like, not particularly at the individual scale (because, as previously noted, few people at the time explicitly compared the two systems or talked about their reasoning), but at an aggregate, social level, across the centuries when the transition was most rapid and evident. I will argue that it was the commonality and ubiquity of the Roman numerals that insulated them against replacement or even any real competition in Western Europe—right up until a set of interrelated disruptions around 1500 made their replacement inevitable.

Three kinds of frequency dependence

Frequency-dependent bias is the term coined by the evolutionary anthropologists Peter Richerson and Robert Boyd to describe cultural transmission in which the frequency of a cultural trait in some population of individuals influences the probability that additional individuals will adopt it (Richerson and Boyd 2005: 120–123). In other words, "a naïve individual uses the frequency of a variant among his models to evaluate the merit of the variant" (Boyd and Richerson 1985: 206). Conformity (a greater-than-expected propensity to conform to the majority) is one form of frequency dependence, but it is not the only form—the conscious choice not to conform is equally frequency-dependent. In frequency-dependent situations, the characteristics of the trait being replicated are potentially irrelevant to its transmission.

Another form of bias that is relevant here is *prestige bias*. While a frequency-dependent bias works on the basis of the popularity of some phenomenon, prestige biases work on the basis of their adoption by important, prestigious, or notable individuals in some context. When some important person—whether the King of France or the latest YouTube "content creator" sensation—adopts a phenomenon, it's not the number of people who are using something but their importance that matters. Both frequency-dependent and prestige biases are thus kinds of indirect bias, as opposed to direct biases, in which the usefulness of the trait for some function(s) is the characteristic being evaluated. If Roman numerals and Western numerals were being directly compared with one another, we might expect a lot of metanotational comparison—people looking at the two systems and giving explicit comparisons. Because that sort of comparison is rare, we should suspect that indirect biases are at play.

Frequency-dependent and prestige biases are two of several important ways in which cultural transmission differs from biological transmission. Biological transmission of genetic material proceeds, metaphorically speaking, in a *vertical* fashion, from ancestors to descendants. In contrast, cultural transmission is both vertical and horizontal—we acquire culture, including (usually) our first language, from parents, but many other aspects of our lives are shaped through our interactions with non-kin, including particularly those of a similar age—our friends, allies, and even our rivals—and through media. When we choose, as youths or adults, to adopt or reject some new innovation, we are mostly talking about horizontal transmission. Western numerals were transmitted through arithmetic texts, in mercantile practice, and through everyday use in inscriptions, books, and manuscripts, and thus fit this model well.

Boyd and Richerson are writing from a perspective in which the adoption and transmission of culturally derived traits motivates long-term and large-scale cultural change. They are not, however, chiefly concerned with individual motivation. Yet, once we take motivation into account—in other words, by adopting specifically cognitive perspectives—it becomes clear that frequency dependence is not a unitary phenomenon. The three examples above illustrate the variety of decision-making processes underlying frequency-dependent outcomes.

In Case A, it may or may not be the case that Whackalot clubs are superior to other golf clubs. The perception that they have a positive utility for playing golf (longer drives, reduced hooks and slices, etc.) is based on the presumption that, if many people are using them, those people must have perceived something beneficial about them. Adopting these clubs is not merely a trend or fad, but reflects an assumption about how others are believed to make decisions about which golf club to adopt. Frequency in this case indexes *perceived utility*. This is related to the phenomenon identified by James Surowiecki as the "wisdom of crowds," where, in a wide sample of a population, an average of individual choices is likely to approximate the best possible answer (Surowiecki 2004). But it draws on a tradition in the psychology of individual and group behavior that is much older—Charles Mackay's 1841 study *Extraordinary Popular Delusions and the Madness of Crowds* showed that numerous hoaxes, follies, economic bubbles, and other trends were based on the same sort of reasoning (Mackay 1963). Surowiecki's point, though, is that in many cases crowd wisdom is correct. Many contemporary systems on the internet, such as upvoting / downvoting on social media or search engine algorithms, depend on the frequency with which previous users have selected some alternative to determine what to show new users. The advantage of this process is that it provides some information to users about what they might prefer; the downside is that such algorithms sediment biases and prejudiced notions about preferences, making them seem neutral (O'Neil 2016; Seaver 2018). Regardless of their positive or negative effects, let's call such cases where adopters assume that a frequent variant is useful by the name *crowd wisdom*.

Case B is similar, but adopters of such features as zipper-pocket jeans, or bell-bottoms, or Capri-length pants, may be fully aware that these items may not be better than what came before them in any absolute sense. They are a trend or a fad, but are nonetheless useful—their utility lies in conforming with that trend in order to acquire social status. Here, frequency measures an item's stylishness. It is socially useful to adopt it: one may hope to achieve wider popularity in doing so. It may even increase one's reproductive success. Nevertheless, there are potential costs. There is the cost of acquiring the trendy item, which may be expected to rise in price if demand outstrips supply. Also, like all trends, it is unlikely to be exceedingly long-lived, and the social advantage to be achieved through a frequency-dependent trait adoption in this case will be fleeting. But the utility is intricately linked to the popularity of the trend in a way that is not true of Whackalot clubs, which may actually be better than less popular alternatives. The study of these sorts of trends has a long history in anthropology, going back at least as far as A. L. Kroeber's (1919) pioneering studies of changes in fashion such as skirt length and seen more recently in Greg Urban's (2010) studies of cultural motion. This work influenced archaeologists who studied frequency seriation, including the famous "battleship curve" whereby new styles in ceramics, or projectile points, or tombstone decoration, or any other aspect of material culture start out as infrequent, growing rapidly to some maximum and then fading away (Dethlefsen and Deetz 1966; Lyman and Harpole 2002). The evaluation of the trait is always dependent on its current popularity, with adopters following the crowd, not because they think the crowd is eternally right, but because the crowd's "rightness" is a social product evaluated at each moment. We might use the term "conformist" to describe this thought process, but terms like "conformity bias" are used in some literature to describe frequency dependence in general, which might be confusing. Bearing in mind that we're really talking about fads and trends here, where what is going on is simply following some newly popular variant, let's call this kind of frequency dependence *faddish*.

Case C has similarities with both A and B, but is substantially more complex. Because Headshot is a communication technology that is closed to nonusers, its perceived utility is not dependent solely on the assumption that anything popular is likely to be superior, nor on the perception that it is socially advantageous to conform to a trend. Rather, because it is used for communication among other users, *its utility is actually dependent on its frequency*. The larger the number of users, the more people any individual user can communicate with. A less popular, noncompatible but otherwise similar system would be simply less useful for the function for which it is designed: communication. In other words, this is a case where a frequency-dependent bias is also a direct bias. Let us call this kind of frequency dependence *networked*, because it is in the exchange of information within a network that it becomes useful.

What these three cases have in common is the feature that in each case, the frequency of something in the environment influences the probability that people will decide to adopt it. Where they differ is in the motivations underlying frequency dependence. But the fact that there are three distinct patterns of frequency-dependent bias is not simply typological trivia. This distinction has major implications for the retention and replacement of cultural phenomena, technologies, and, as we will see, numerical notations.

In crowd-wisdom frequency dependence, the expectation of utility of some cultural phenomenon for some function may be followed up with an evaluation of its utility. If it turns out that Whackalot clubs were popular for some other reason than their overwhelming superiority (perhaps due to a prestige bias—some famous golfer uses them), they may be abandoned once this becomes clear. Even if they are superior for golfing, unless these clubs are the best clubs conceivable, at some point a new set of clubs will be developed that are even better, at which point a new cycle of frequency-dependent bias, leading to adoption, leading to evaluation (and eventual abandonment) may arise. But remember that each new technology, no matter how good it is, starts from a frequency of zero—no one is using it when the product is first launched. Frequency dependence is something to be overcome by users perceiving the new technology to be better—in other words, it doesn't merely need to "be" better; it needs to be recognized as such, indeed as sufficiently superior to warrant replacing the old technology. The frequency curves of three hypothetical technologies under crowd-wisdom frequency dependence might look a little bit like figure 4.1. The popularity of any one variant (on the graph, its maximum) lasts only as long as there is nothing that is perceived as sufficiently superior to replace it.

In faddish frequency dependence, however, the utility of a style of clothing cannot be evaluated adequately independently of its popularity. What makes a fashion fashionable is that it is in fashion. There may be other features than frequency dependence involved (prestige bias, cost, etc.), but because its popularity is a primary determinant of its perceived utility, when it fades as a trend, as it surely will, it is very likely to be replaced, sometimes quite rapidly. Nothing has to be "bad" about the old fashion—the mere presence of some new fashion may be enough to replace it. Some trends, like colors or skirt lengths or jean leg widths, may be cyclical and may fall in and out of fashion multiple times—this is what Kroeber found about many of the trends in women's fashion he examined. Others, like the 1920s fad of pole sitting, may come into vogue once and then never be heard from again. But why do trends fade at all? Simply put, because people like new things, and old fashions lack appeal. So this is a case where frequency dependence may be counteracted by novelty-seeking behavior by individuals seeking to get ahead of trends. The frequency curve for three hypothetical phenomena under conditions of faddish frequency dependence might look a little bit like figure 4.2.

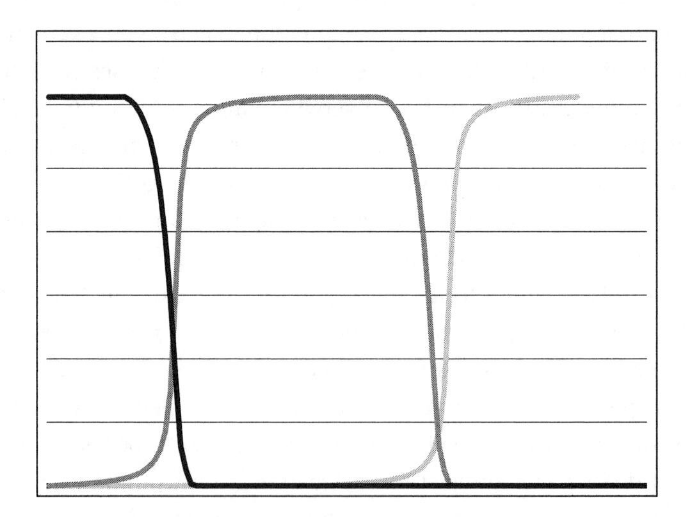

Figure 4.1
Crowd-wisdom frequency dependence

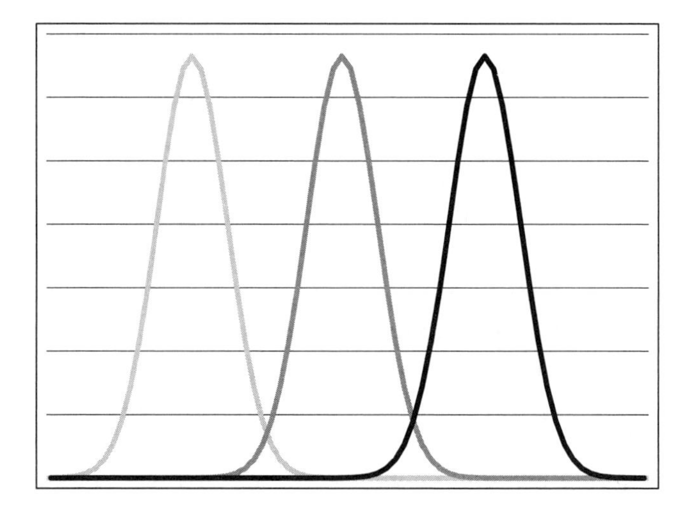

Figure 4.2:
Faddish frequency dependence

Networked frequency dependence combines features of both of these. Like items favored by crowd wisdom, a communication technology can have intrinsic features that promote or limit its popularity. So, for instance, a social media platform may decide to start bombarding its users with unwanted advertising, or may introduce content restrictions that are seen as undesirable, leading to its abandonment. As with faddish frequency dependence, because one of the criteria that is used to evaluate a communication technology is its popularity, its utility cannot be evaluated fully independently of its popularity. But in this case there is an additional cost to abandoning the technology, namely that one will be cut off from communication with other users. Abandoning something popular means choosing to abandon the network still engaged with it. As a result, we might expect that such technologies will be longer-lived than fads, and will also be evaluated in terms of their utility for some function. We might also expect that in many circumstances, it becomes much more difficult for a competitor to break into some domain of activity—that it would be relatively rare for a new variant to take over from the old one completely, and that relative stasis would be the normal condition within any particular community or context. New innovations might arise but then be rapidly discarded, not because they are poor in some absolute sense, but because they are rare initially and are unable to overcome the inertia of an extremely popular system already in existence. This might look a bit like figure 4.3.

In addition to patterns of replacement, a further difference between these three types of frequency-dependent phenomena involves how faithful replication is likely to be. The characteristics of networked frequency dependence encourage extremely faithful phenotypic replication of traits being emulated. Because one must be understood for the trait to be of any value, one must communicate very similarly to one's peers. Writing systems are good examples of communicative systems whose stability is encouraged by the need to communicate with others—stray far, and you will not be understood. Change does happen, but it is slow. With faddish cases, on the other hand, while some fidelity is necessary to enjoy the social benefits of conformity, there may also be some advantage to a slight but significant alteration of the phenotype in order to "get ahead of the curve" of the trend. With phenomena whose utility is partly independent of frequency (such as traits influenced by crowd wisdom), users might make adjustments to their behavior in order to gain an advantage over the many other users of the same trait. The first two types lead to further innovations (and in faddish frequency dependence, at least, this can lead to the trend failing), while networked frequency dependence tends to lead toward the long-term retention of a single variant within a social system.

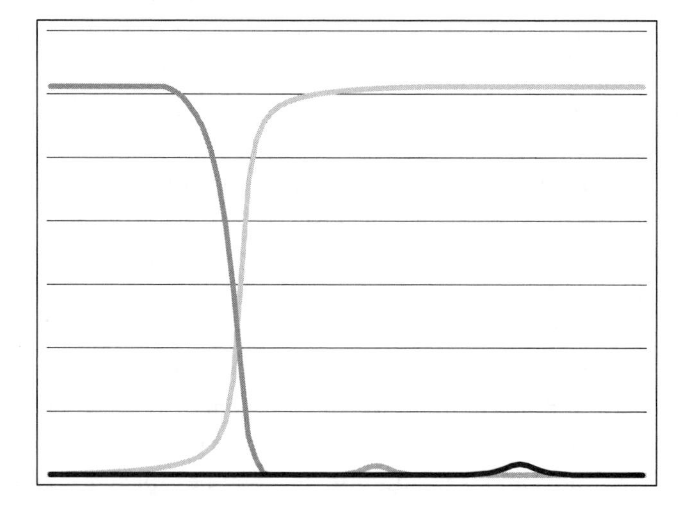

Figure 4.3
Networked frequency dependence

The question we then need to ask is: In conditions where networked frequency dependence applies, why and how do new innovations come to replace old ones at all? If you've spent any time around social media and chat programs, you may have noticed that, far from being shaped by persistent, long-lived phenomena, interactive electronic communication over the past thirty years has transitioned from bulletin board systems and IRC chat to Usenet and AOL Instant Messenger to Livejournal, Myspace, Friendster, Facebook, Twitter, Instagram, Reddit, Tumblr, Snapchat, and more. How did each of these overcome the popularity of the preceding technology? In part, the development of new features made a difference—for instance, many users of chat programs appreciated the feature whereby the program tells you when your interlocutor is typing, prior to their message coming through (Machak 2014). But speak to nearly any user of Facebook today and you will not hear from them that they believe it to be technologically superior. The choice of technology is deeply influenced by the network of users with whom one can be put in contact through using it, and the factors influencing any particular network's choice may be different (Kwon, Park, and Kim 2014).

In the early years, prior to the late 1990s, internet access was rare enough and diffused across numerous enough platforms (Usenet, AOL, IRC, etc.) that there was no "killer app." Most people were becoming users of social media for the first time, and this was true right up until the middle 2000s, when many of the now-established platforms were founded. One might then expect that whichever site won out—in this case, perhaps Facebook and Twitter—would dominate all the others to such a degree that no competition would be possible. However, there is another important source of new users—young people, who as they mature into their teenage years pick social media not on the basis of the total number of users of some platform, but the number of users relevant to them—i.e., largely their peers (Hofstra, Corten, and Van Tubergen 2016). Adoption of Facebook among high school students is now substantially lower than it was ten years ago, because of the perception that it is now an adult-oriented medium where their parents and other adults can surveil their activities. In other words, its popularity among one set of users actively reduces the likelihood of its adoption among another—most teens do not have an overwhelming desire to communicate online with their elders. As a result, although the largest social media platforms do exercise some frequency-dependent dominance among some groups—it is unlikely that Facebook is going to collapse anytime soon—the constant infusion of new users of social media whose goals and interests may be very different from their predecessors' ensures fluidity in this domain.

Networks and frequency in communication systems

Networked frequency dependence is extremely commonplace in communication systems. Many approaches, including several of those we saw earlier, presume that people are more likely to directly compare the utility of two technologies or phenomena than to depend on indirect biases, but this presumption must be tested in each case. Did VHS video technology succeed over Beta because it was cheaper (despite Beta's higher image quality)—a selectionist argument? Alternatively, after achieving an initial level of success (frequency), did VHS win simply because once more video stores stocked VHS and more distributors produced VHS products, more viewers were inclined to purchase the most popular communication technology? Brian Arthur (1990) was among many favoring the latter explanation, although there is a complex relationship between these sorts of feedback effects, economic systems including pricing, and other social factors. To choose a more modern example, are Windows operating systems actually fitter than Mac or Linux ones, or is their success due to the fact that the choice of operating system can affect one's ability to share files or to use a colleague's computer while away from home? Movement in the direction of computers that can read or convert between formats may have mitigated the frequency-dependent effects in this case, allowing users to make choices based on personal preference rather than popularity.

In an interesting study, Alex Bentley (2006) uses data derived from keyword searches in the ISI Web of Knowledge database (one of the principal databases of scholarly articles) to critique sloppy language use in archaeology, using an evolutionary anthropological framework. Using data on the words "nuanced" and "agency" used in titles, abstracts, and keywords of articles, he suggests that these terms are used by humanistic archaeologists in a way analogous to faddish frequency dependence—as "buzzwords" intended to impress others. Under this theory, their increasing frequency reflects conformity to ongoing trends as a status-seeking mechanism, but otherwise lacks any intellectual purpose. In responding to Bentley, however, I have shown that the use of "agency" as a keyword is instead a social and communicative strategy adopted by authors who wish their work to be found and read by others (Chrisomalis 2007). When you write an abstract or title, or choose keywords, you are not merely signaling to gain status, but to be found and read. Thus this case is much more like networked frequency dependence; keywords become popular because authors and readers have an interest in using shared communicative conventions when using keyword searching in electronic databases like the Web of Knowledge. If authors used phrases like "the theory that individual action influences large-scale processes" in place of "agency," their work would be less accessible, less widely read, and less frequently cited. Keyword searching is a communicative technology in which the use of common terminology allows author and reader to find one another more readily, and in which a frequency-dependent bias leads to certain keywords' popularity increasing on that basis.

As early as the 1930s, linguists such as George Zipf noted frequency-dependent effects in linguistic transmission, as we saw in chapter 2 (Zipf 1935). This work was later fundamental to Joseph Greenberg's development of the concept of markedness, establishing that unmarked linguistic forms were virtually always more frequent than marked ones (Greenberg 1966). These accounts differ significantly from that of Boyd and Richerson in that they adopt an explicitly structuralist model for the organization of knowledge, but like them assume a relatively naïve transmission process in which the cognition of the recipient is not necessarily relevant to cultural replication. They are also studies of outcomes, rather than of cognitive decision-making processes underlying those outcomes.

The study of frequency-dependent linguistic phenomena thus remains incipient, despite these early efforts. Lieberman et al.'s (2007) research linking lexical frequency to the regularization of English irregular verbs is a notable exception. This study argued that, among irregular verbs, those more commonly used were less likely to undergo regularization. Why should this be? From a purely efficiency-based perspective, the opposite should be true—people should seek to simplify that which is common, where the pressure would be greatest to make some improvement. The answer they provide is cognitive, relying on the limits of human memory for rarely encountered phenomena. For frequent irregular verbs like *be* and *go*, constant exposure to verb forms such as *was* and *went* continually reinforces their use. You use those verbs every day and are unlikely to say **goed* unless you are a "new user"—e.g., a child, overregularizing based on a pattern not yet fully established in your mind. In contrast, for a rare irregular verb like *stride*, English speakers are less certain about *strode* vs. *strided*, which most of us use rarely if at all. Analogies with other, more common verbs don't help us much here, because we have *divide/divided, hide/hid,* and *ride/rode,* any of which could be the model we choose. So, when in doubt, we turn to the regular morpheme *-ed*. Over time, **strided* may be predicted to become more common and eventually replace *strode*.

In the domain of numerals, Coupland's (2011) corpus-based analysis of the frequency of English numeral words in online usage, which I have discussed briefly earlier, shows how spoken and written number words may be affected by frequency dependence. Coupland starts with the recognition that lower numerals are more frequent than higher ones, although that fact is not in itself a product of frequency dependence. Some of the other factors he establishes, however, such as the greater frequency of certain round numbers, or of "geminated" numerals with repeated elements, like *ninety-nine*, might well be a product of long-established patterns of networked frequency dependence. His hypothesis that humans prefer scale marking using small quantities, for instance preferring *ten minutes* to *six hundred seconds*, is also amenable to a frequency-dependent interpretation. We are only at the beginning of this sort of study.

This is not to say that frequency dependence is the sole factor behind the adoption of communication technologies. Stephen Houston (2004) shows that, among writing systems in ancient states, the goal was frequently not to be understood by large numbers of nonelites, but rather to ideologically manipulate others by impressing the difference between those who possessed the skill and those who did not. In some cases, texts may not even have been intended for human eyes, but solely for those of the gods: here the production of such texts serves to highlight the writers' rare ability to communicate with those entities. Similarly, writing can serve cryptographic or exclusionary functions, to be understood among a select group of individuals but no further. Literacy rates in most of the premodern world were extremely low, and literate communities were limited in size. This does not negate the general pattern that communication technologies are used for communication, and as such we ought to be aware of situations where utility for communication is evaluated by potential users in terms of the size of the networks involved.

Roman and Western numerals: A case study in frequency dependence

We saw in the previous chapter that many histories of mathematics presume that numerical notation systems are adopted on the basis of their (perceived or actual) superiority for performing arithmetic (Dehaene 1997:101; Ifrah 1998: 592)—that is, through a direct bias that does not involve considerations of frequency dependence. However, we also saw that written numerals historically have rarely been used directly for computation; written numerals are used primarily to record and transmit numerical information, while other techniques are used for performing computations. Numerical notation is thus primarily a communication technology rather than a computational technology. Because such notation is relatively easy to learn (low initial cost) and because it is translinguistic (thus not dependent on knowledge of an entirely different and complex symbol system, i.e., a language), it makes a useful case study in frequency dependence.

Over the past 1,500 years, a variety of historically related decimal, place value systems with signs for 1–9 and 0 have been used, first in India and spreading from there throughout the world. The Western numerals, which have several billion users, and the Tibetan numerals, which have a few hundred thousand users at best, are structurally identical notations. As communication systems, however, the Western numerals are generally preferred because they allow one to communicate with many more individuals than would be possible with any other system. Even though the cost of learning a new system is low—ten symbols are not particularly difficult to memorize—Western numerals persist in part because they are so popular. Tibetan numerals, on the other hand, are limited both by a frequency-dependent bias *and* by a prestige bias in favor of the Chinese numerals, which have hundreds of millions of users, as well as significant social and political constraints on their spread since the 1959 occupation of Tibet. For most of the less common ciphered-positional decimal systems—e.g., Tibetan, Lao, Khmer, Javanese, Bengali, Oriya—the main system replacing them is the Western numerals, a structurally identical system. Here, frequency-dependent and prestige biases can really be the only explanation.

But we need to start earlier, with the Roman numerals, which for over a millennium were the only system in use throughout much of Western Europe. They were the frequent variant, not the rarity. For the most part, the Roman numerals persisted in late antiquity and the early medieval period despite the presence of other systems that could have competed with them—Greek alphabetic numerals to the east, for instance. It would have been very easy for medieval scribes to replace the letters of the Greek alphabet with Roman letters, and in fact there are a handful of texts that use Latin alphabetic numerals (a = 1, b = 2, c = 3 ... k = 10, l = 20 ... t = 100, u = 200 ...) (Lemay 2000; Burnett 2000). Such numerals show up in a handful of translations of astronomical texts that use some other alphabetic numeral system (like Greek or Arabic) starting in the twelfth century, then disappear entirely in the thirteenth. At the time, virtually everyone in Western Europe—even mathematicians—was using Roman numerals, and so the new system never caught on. The Western numerals were largely restricted to a set of mathematical texts, and even then to particular sorts of arithmetical work in those texts.

The frequency-dependent bias in favor of the Roman numerals involved not only the frequency of present users but also of past users, since the ability to use and read older texts depended on knowing them. In an era when the transmission of written knowledge was dependent on scribes reading and writing old texts, faithful replication of Roman numerals was actually quite important. Where the Roman numerals had changed, this could cause problems, as in the texts discussed by Shipley (1902) where the ninth-century scribes did not understand fifth-century manuscripts and thus made systematic errors. For instance, because XL was the most typical form for 40 in the ninth century whereas XXXX had been quite common in the fifth century, the later scribes often presumed that the fourth X was an error, and so deleted it. These and other errors are only minor blips in what is otherwise a highly persistent, extremely common notational practice across space and time. This motivation for retention, to be able to read old materials, is quite different from, say, motivations in connection with social media, where the ability to read and use decades-old material may not be socially valued. It does have a rough parallel with technologies like VHS—long supplanted by DVD, Blu-Ray, and now by streaming video—which is why many of us have a box full of old VHS tapes of holiday concerts from years gone by, and a dusty VCR hanging out in storage somewhere, probably never to be used again.

The Western numerals themselves were once a newfangled, foreign innovation used mainly by infidels—that is, there was every social reason imaginable to reject them in favor of the tried and tested Roman numerals (with the handy abacus for computation). Their eventual predominance was not a rapid and straightforward process of replacement; rather, the Roman numerals persisted for centuries as the notation of choice for scribes, merchants, and administrators. In some cases, the use of Western numerals was discouraged, ostensibly because their unpopularity led to the potential for obfuscation and fraud (an *infrequency*-dependent bias!) (Struik 1968).

In the late tenth century, Gerbert of Aurillac (who would later become Pope Sylvester II) introduced the use of the "abacus with apices" into Western Europe. Essentially this was a counting board on which counters or jetons were placed; not unmarked counters as on a Greco-Roman abacus, but instead notated with one of the nine *apices*—the Western digits 1 through 9 (Burnett and Ryan 1988). The user could then place these on a grid or table to manipulate numbers in a materialized form. This is not a cumulative-positional abacus using the repetition of objects (beads, pebbles, etc.), but a ciphered-positional abacus where the counters serve in lieu of writing down calculations. But this was a competitor for the abacus, not for the Roman numerals, which could have been written as apices on a token just as easily as Western ones could. As we see in figure 4.4, a manuscript depiction

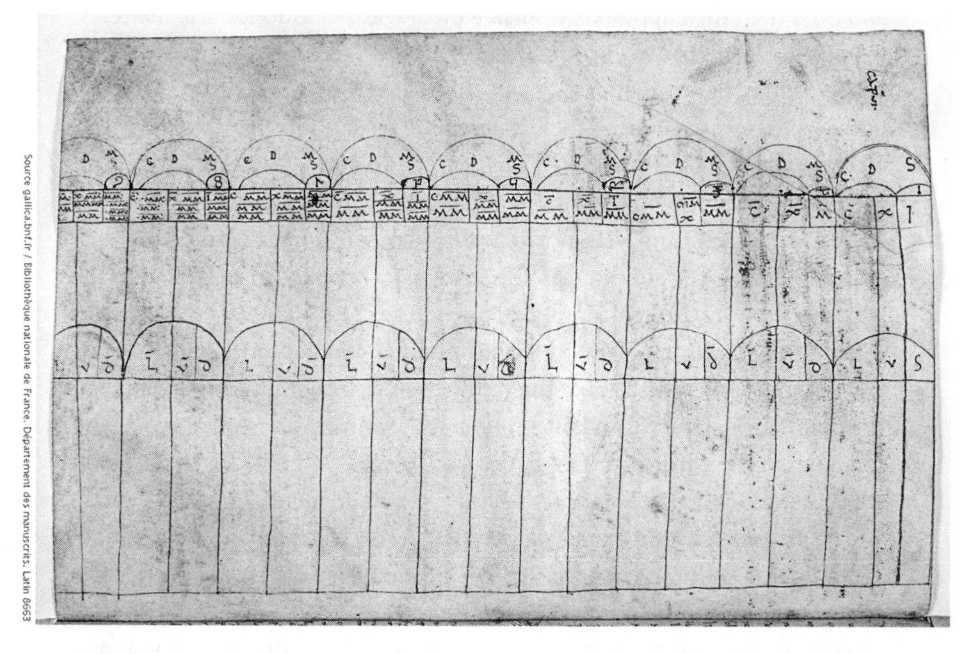

Figure 4.4
Abacus with apices (source: Bibliothèque nationale de France, Latin 8663, folio 49v)

of the abacus with apices, the columns could readily be notated in Roman numerals, even as the Western numerals began to be used.

As we saw in the previous chapter, the choice between the use of the reckoning board (medieval abacus) and the Western numerals and written arithmetic was complex and rarely a matter of selecting one notation over the other. Leonardo of Pisa's *Liber abaci* promoted the use of written arithmetic with place value numerals but was never intended to supplant Roman numerals. Later medieval mathematicians advocating for what we can, in an overly blunt fashion, call the "algorist" position continued to use Roman numerals in all sorts of ordinary contexts, including when writing about mathematical subjects. Just as al-Uqlidisi recommended using Arabic numerals only for doing the computation but some other system (such as the abjad numerals) for writing down answers, Western Europeans were simply not concerned with possible deficiencies of the Roman numerals and were happy to keep using them as an everyday notation, even despite the availability of an alternative. At this point, the use of Western numerals was still extraordinarily rare outside of a narrow group of specialists. Only in the thirteenth century and more in the fourteenth did Western numerals show up among other users, and even then their appearance outside theological or astronomical matter is remarkable at this period. So, for instance, the Genoese notary Lanfranco used Western numerals in the margins of his cartularies between 1202 and 1226, but usually only to indicate the honoraria he received for his work (Krueger 1977). Some thirteenth-century sculpted figures at Wells Cathedral bear mason's marks with Western numerals to indicate their correct position, but the cathedral accounts were notated using Roman numerals until the seventeenth century (Wardley and White 2003: 7). Pritchard (1967: 62–63) discusses an early fourteenth-century graffito at the church at Westley Waterless, in Cambridgeshire, enumerating grapevines in a series of rough-carved lines, and suggesting, on the basis of the rarity of Western numerals in England at the time, the presence of an Italian priest at the church. These represent thin, sporadic evidence, barely a trace in comparison to the thousands upon thousands of inscriptions and manuscripts with Roman numerals from this period.

The fifteenth century marks, roughly, the period when the Western numerals began, in large parts of Europe, to be well enough known that their appearance is no longer remarkable in most texts. Carvalho (1957) examined a variety of late medieval text genres in Portuguese, showing that it was not until around 1490 that the Western numerals became predominant in travel accounts and scientific documents, and still later for other contexts. Crossley's

(2013) examination of catalogues of dated manuscripts showed an increase from 7% of dates in Western numerals in the thirteenth century to 47% by the late fifteenth century. In the Catasto tax records and fiscal documents of Florence from 1427, Rebecca Emigh shows that Western numerals were used not only by central authorities but also by local households (Emigh 2002). Still, on the whole, the Roman numerals were substantially more common throughout the entire fifteenth century across all of Europe than the Western numerals.

The sixteenth century marks the major inflection point, the time at which the Western numerals became, if not ubiquitous yet, clearly preferred for most functions in most areas. England, which in many aspects was the slowest to adopt the change, certainly did so enthusiastically by the period from 1570 to 1630, as shown by two studies involving English probate inventories: Wardley and White's "Arithmeticke Project" (2003) and Cheryl Periton's recent doctoral dissertation (2017). Probate inventories are inventories of goods, money, and property owned by a person at the time of their death. They are particularly useful because they are so widespread. Wardley and White developed and used a taxonomy for identifying texts using Roman numerals alone, Western numerals alone, Roman numerals but Western totals, Western numerals but Roman totals, mixed notations, lexical (number words), and some miscellaneous categories. This fine-grained approach allows a year-by-year demonstration of the shift. Periton (2017), similarly, uses this model to discuss the prevalence of Roman numerals, Western numerals, and English number words in a small corpus of 92 probate inventories dating from 1590 to 1630, showing that while there are surely trends toward writing dates in Western numerals, lexical and Roman numeral representations continued throughout this period.

There were still holdouts—for instance, Sayce (1966) shows that Parisian bookbinders were still using Roman numerals internally, for markings on books designed for industrial practice, throughout the seventeenth century and even up until 1780! William Cecil, Lord Burghley (1520–1598), the Lord High Treasurer of England under Elizabeth I and the figure responsible for most English economic policy, is said to have translated documents written in Western numerals back into Roman numerals, with which he was more comfortable (Stone 1949: 31). At the level of individual choices, this is to be expected. As an aggregate, the Western numerals had secured their position by this time.

All of the above provides some evidence of a roughly S-shaped curve of the frequency of the Western numerals, starting out very slowly and rising rapidly until, by the late seventeenth century, writers such as John Aubrey could lament, and John Wallis could praise, the fact that no one really used Roman numerals any longer. But what this sort of data doesn't tell us is the motivations of users for switching. And given how little metalinguistic or metanotational material directly bears on the evaluation of the Roman numerals, it's hard to know whether and for what purposes anyone found the Western numerals to be useful. For this process to get started, the Western numerals had to come from a position of radical *in*frequency, and the process of replacing Roman numerals took centuries to complete. This was one of the reasons originally given by their detractors for not adopting them: they were obscure, no one knew them, so they were prone to falsification and fraud. How did the ubiquitous Roman numerals come to occupy the position they now have, as an archaism and a bogeyman of the middle-school classroom, after their medieval ubiquity?

The answer, I think, lies in another radical change in communication technology: the printing press. As Elizabeth Eisenstein (1979) and others have argued, the printing press was an enormous agent of social change, and one of the most important of its effects was to open up literacy to a much wider range of users—the burgeoning European middle class—than had ever been possible in a manuscript-based literary tradition. Printers became the vectors through which a centuries-old notation, the Western numerals, could become accessible to far broader audiences. Newly literate people, unencumbered by the tradition of Roman numeral use, and unconcerned with what system had been used in the past, adopted the Western numerals. These newly literate individuals, reading and writing outside the medieval manuscript tradition and engaging with printed texts, were the disruptive force that put an end to the dominance of the Roman numerals in Western Europe. Jessica Otis (2017) argues convincingly, based on textual analysis of English arithmetic books and their marginal notes, that the spread of printed arithmetic texts through the sixteenth and seventeenth centuries brought literacy and numeracy together for the first time, at least in England. We are used to thinking of literacy and numeracy as going hand in hand, linked to formal schooling and in fact to preschool activities, but this is a modern idea, a product of a particular historical moment.

This is not to say that the adoption of the Western numerals was instant upon the adoption of printing. One strategy for examining the transition in a more fine-grained way is to look at texts that use both Western and Roman numerals but for different purposes. Books that use only one system may indicate a printer's preference for one system or another, or simply that the audience for a particular text was expected to know only one or the other. In contrast, books that use both systems suggest the functions for which each system was perceived as useful, and thus give us a window into the motivations of the printers relative to their audiences. The corpus of early English incunabula (the technical term for printed books published before 1501, during the earliest decades of European movable-type printing) as well as early sixteenth-century printed books provides a robust sample. Searching through the full-text Early English Books Online database, I found 55 books dating from between the beginning of the English printing tradition and 1534 that contain both Western and Roman numerals. If Western numerals were adopted in part because of their utility for foliation (page numbering), chapter headings, or indexing, then one might expect that they would most often serve organizational functions, while Roman numerals might be retained in colophons or title pages, perhaps for recording publication dates. If no such pattern exists, then the organization of texts for the convenience of readers was probably unrelated to the choice of numerical notation. This analysis will help us better understand how printers were thinking about the new notation, its advantages, and its weaknesses.

Table 4.1 lists the major organizational features of books that use numerals, along with the number of books in the corpus of 55 books that use a particular system for that function. For most functions, Roman numerals are preferred throughout this period, the exceptions being for title pages (a late development) and marginal notes (which are rare in the earliest decades of the period under study).

Table 4.1

Roman numerals and Western numerals in early English printed books, 1470–1534

Feature	Roman	Western
Colophon	25	21
Title page	0	6
Foliation	26	17
Table of contents	7	7
Marginal notes	2	6
Chapter headings	11	2
Signature marks	29	14

Western numerals were rare in England before 1530, in both manuscripts and printed texts (Jenkinson 1926). William Caxton briefly experimented with Western numerals in six of his works printed at Westminster between 1481 and 1483 (Blades 1882: 47). However, Caxton used Western numerals only for signature marks: organizational features used by bookbinders to ensure that pages are placed in the correct order, but not really intended for the eventual reader at all. He printed all the other organizational features, such as chapter headings, in Roman numerals. This suggests that while he expected (or knew) that his bookbinders were familiar with the new system, it was still novel enough to his readership that he did not employ them in features intended for them. Figure 4.5 shows a leaf from Caxton's *Reynard* of 1481 (STC 20919) containing the first Western numeral ever printed in an English book (the signature mark "a2," bottom right), alongside the chapter headings in the table, numbered in Roman numerals. Caxton's experiment was short-lived; from 1484 onward, he reverted to Roman numerals for signature marks and used no Western numerals anywhere else. Indeed, these were the only Western numerals in English incunabula.

The use of Western numerals in books printed in England resumed only in 1505 with the printing of Hieronymus de Sancto Marco's *Opusculum* (STC 13432), probably by Richard Pynson in London. The choice to use Western numerals was more than simply one of picking a notation, since, to use Western numerals in print, printers needed to develop fonts that included the new characters. Although the foliation (page numbers), signature marks, and table of contents all use Roman numerals, the date that appears in the bottom line of page xxix is in highly irregular Western numerals (figure 4.6). Each numeral comes from elsewhere in Pynson's type set: the 1 from majuscule I, the 0 and 5 from black-letter o and h, and the 9 (in the day number) from the apostrophe used for the Latin syllabic ending *-us*. Very likely the printer lacked a set of Western numerals, which may explain why they were not used throughout the text. But in that case, why were they used at all? And note that the volume is still paginated in Roman numerals—how could it be otherwise, if the printer had no numerals in his type set? This is, bear in mind, at least three centuries after the Western numerals were first used in England, and five centuries after their first use in the Latin mathematical tradition.

Throughout the corpus I examined, 16 of the 55 books examined are, like the *Opusculum*, dated in Western numerals (either in the colophon or

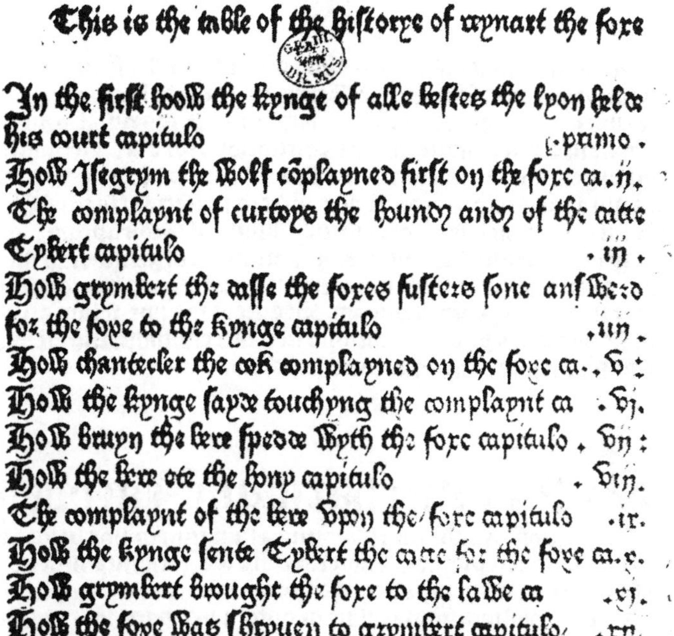

This is the table of the historye of reynart the foxe

In the first book the kynge of alle bestes the lyon helde
his court capitulo .primo.
How Isegrym the wolf complayned first on the foxe ca. ii.
The complaynt of curtoys the hounde and of the catte
Tybert capitulo .iii.
How grymbert the dasse the foxes susters sone answerd
for the foxe to the kynge capitulo .iiii.
How chanteclere the cok complayned on the foxe ca. .v.
How the kynge sayde touchyng the complaynt ca .vi.
How bruyn the bere spedde wyth the foxe capitulo .vii.
How the bere ete the hony capitulo .viii.
The complaynt of the bere vpon the foxe capitulo .ix.
How the kynge sente Tybert the catte for the foxe ca. x.
How grymbert brought the foxe to the lawe ca .xi.
How the foxe was shryuen to grymbert capitulo .xii.
How the foxe cam to the court & excused hym ca .xiii.
How the foxe was arestid and Juged to deth ca .xiiii.
How the foxe was ledde to the galwes capitulo .xv.
How the foxe made open confession to fore the kynge &
to fore alle them that wolde here it capitulo .xvi
How the foxe brought them in daunger that wolde
haue brought hym to deth And how he gate the grace
of the kynge capitulo .xvii.
How the wulf and the bere were arestyd by the labour
of the foxe capitulo .xviii.
How the wulf & his wyf suffred her showes to be pluckyd
of And how the foxe dyde them on his feet for to

a 2

Figure 4.5

Leaf from Caxton's *Reynard* of 1481—with "a2," bottom right, the first printed Western numeral in an English book (STC 20919)

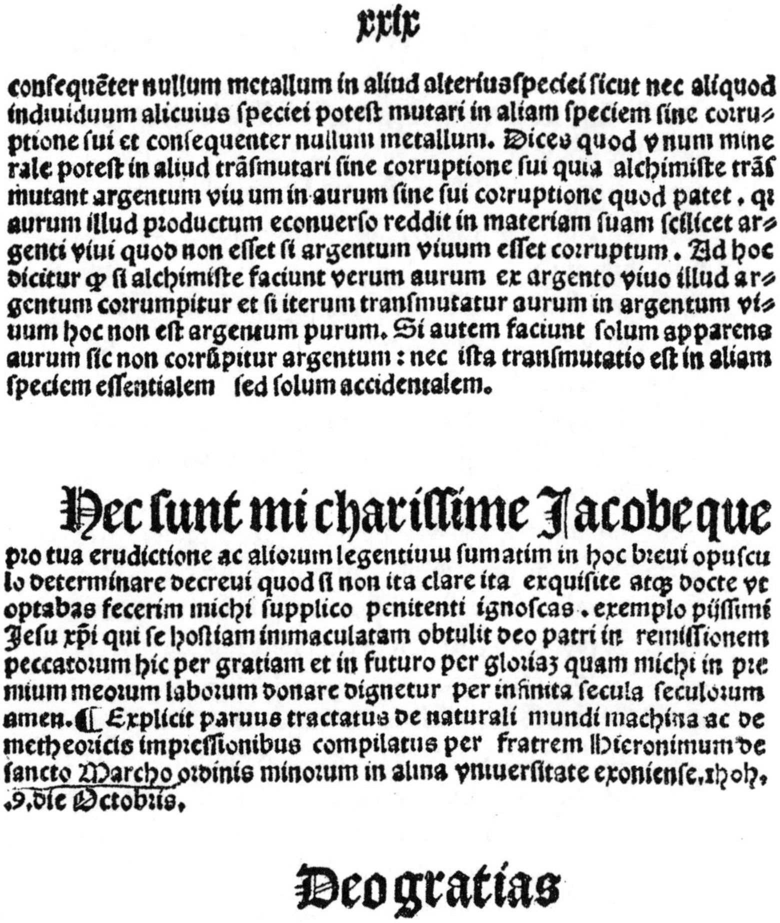

xxix

consequēter nullum metallum in aliud alterius speciei sicut nec aliquod indiuiduum alicuius speciei potest mutari in aliam speciem sine corru-ptione sui et consequenter nullum metallum. Dices quod vnum mine rale potest in aliud trāsmutari sine corruptione sui quia alchimiste trās mutant argentum viu um in aurum sine sui corruptione quod patet. q2 aurum illud productum econuerso reddit in materiam suam scilicet ar-genti viui quod non esset si argentum viuum esset corruptum. Ad hoc dicitur q2 si alchimiste faciunt verum aurum ex argento viuo illud ar-gentum corrumpitur et si iterum transmutatur aurum in argentum vi-uum hoc non est argentum purum. Si autem faciunt solum apparens aurum sic non corrūpitur argentum: nec ista transmutatio est in aliam speciem essentialem sed solum accidentalem.

Hec sunt mi charissime Jacobeque pro tua erudictione ac aliorum legentium sumatim in hoc breui opuscu lo determinare decreui quod si non ita clare ita exquisite atq3 docte vt optabas fecerim michi supplico penitenti ignoscas. exemplo piissimi Jesu xp̄i qui se hostiam immaculatam obtulit deo patri in remissionem peccatorum hic per gratiam et in futuro per gloriā quam michi in pre mium meorum laborum donare dignetur per infinita secula seculorum amen. ¶ Explicit paruus tractatus de naturali mundi machina ac de metheoricis impressionibus compilatus per fratrem Hieronimum de sancto Marcho ordinis minorum in alma vniuersitate exoniense. xhoh. .9. die Octobris.

Deo gratias

Figure 4.6
Hieronymus de Sancto Marco, *Opusculum*, fol. xxix, printed by Richard Pynson in 1505 (STC 13432)

on the title page) but have folio numbers in Roman numerals. Conversely, 13 books are dated in Roman numerals but foliated in Western numerals. This suggests that, far from presenting an obvious cognitive convenience for organizing books, both Western and Roman numerals were seen as perfectly adequate for either function. Even more remarkably, eight of the 55 books with both Western and Roman numerals use different notations for the foliation of leaves than for the folio numbers in the table or index of the book—four with the foliation in Roman, four in Western. In other words, the reader would look up the folio number in the index in one notation, and then find the correct folio in the book using the other one! While one interpretation of these irregularities is that printers expected readers to be so familiar with both systems that they could rapidly transliterate between them, this can hardly be correct, given the scarcity of Western numerals at this time. What we more likely have is a set of norms still in flux, developing along with its new audiences, with few settled production standards.

By the second quarter of the sixteenth century, the Western numerals were becoming more common in English printed books. Pynson's 1523 edition of Villa Sancta's *Problema indulgentiarum* (STC 24729) was the first book printed in England to contain only Western numerals—no Roman numerals at all. The London printer John Rastell was one of the earliest English advocates for the new system, and in the late 1520s he wrote and printed *The Book of the New Cards* (STC 3356.3), which survives only in a half-sheet quarto including the table, but whose contents indicated how "to rede all nom[bers] of algorysme" (Devereux 1999: 129). Yet Rastell (and his son William, who succeeded him in the early 1530s) never used Western numerals exclusively in any of his books, generally using them only in title pages and colophons, and only sporadically in signatures and elsewhere. Similarly, William Tyndale's Bible of 1535 contains no Western numerals whatsoever—chapters, marginal notes, signatures, and the colophon are all in Roman numerals.

Over the course of the early English printing tradition, a sporadic and inconsistent increase is evident in the use of Western numerals, one that is more evident in colophons and title pages than in organizational features. Their placement in these visually prominent contexts illustrated the printer's familiarity with the new system, without requiring readers to be fluent with the innovation. Ironically, then, the Western numerals served a prestige function in early English printed books for their novelty, just as Roman numerals would eventually serve a prestige function in title pages for their antiquity.

Even printers very favorable to the use of Western numerals in general made frequent use of Roman numerals in organizational contexts such as foliation, side notes, and chapter headings. Inconsistencies in printers' practice suggest that, far from a cognitive revolution, the introduction of Western numerals resulted in some potential for awkwardness and confusion. Sixteenth-century Bibles and prayer books mixed Roman and Western numerals in ways that Williams (1997: 7) describes as "schizophrenic"—at the very least, unpredictably and without any clear underlying pattern. But the mixing of systems in different contexts need not be seen as a retrograde step. Rather, in some cases it served the very useful function of distinguishing parallel enumerated lists—just as even today prefatory material in books, or nested sublists, use Roman numerals alongside Western ones. The revolutionary idea that multiple notations can coexist in parallel for the benefit of the readers represents the most important insight to take from the tradition of early printed books from England. But this is hardly grounds for replacing Roman numerals, but rather, for retaining them.

This hodgepodge of different notations in the first century of the printing tradition puts to rest the claim that Western numeration was responsible for effecting significant cognitive changes in how people organize information. Elizabeth Eisenstein, the historian of printing, claimed:

> Increasing familiarity with regularly numbered pages, punctuation marks, section breaks, running heads, indices, and so forth, helped to reorder the thought of *all* readers, whatever their profession or craft. Hence countless activities were subjected to a new "esprit de système." The use of Arabic numbers for pagination suggests how the most inconspicuous innovation could have weighty consequences—in this case, more accurate indexing, annotation, and cross-referencing resulted. (Eisenstein 1979: 105–106)

In the same year, Rouse and Rouse made the comparison with Roman numerals more explicit:

> Tools of reference required a system of symbols with which to designate either portions of the text (book, chapter) or portions of the codex (folio, opening, column, line). Toolmakers were nearly unanimous in their rejection of roman numerals as being too clumsy for this purpose. Before widespread acquaintance with Arabic numerals, the lack of a feasible sequence of symbols posed serious problems in the creation of a reference apparatus. (Rouse and Rouse 1979: 32)

If this were true, we would expect the data to look very different, with systematic replacements and consistent notation within particular volumes and the printers who published them. It would also imply that crowd-wisdom frequency dependence is more likely than networked frequency dependence as an explanation for the shift—in other words, that the Western numerals really were better, and their adoption was merely a confirmation that users found them useful for organizing information.

But this is not what we see at all. The place where readers would be most likely to encounter the numerals, printed books, was a new medium with only gradually evolving standards and norms for notations. Roman numerals remained common throughout the first decades of the printing tradition and, when abandoned, were abandoned sporadically and without apparent reason. Thus, just as the Western numerals were preferred for reasons other than their arithmetical efficiency, they equally were not the leading edge of a cognitive revolution in organizing and structuring knowledge. Rather, gradually, they developed a critical mass of use alongside a critical mass of new users in the sixteenth century especially (in France and Italy, a little earlier), until their frequency cemented their position as the dominant notation of the region.

Even so, both perceived utility (crowd wisdom) and conformity (faddishness) may still have played a role in the increasing prevalence of Western numerals. In contrast to the Roman numerals, Western numerals were used for pen-and-paper computation, and the perception that they were better for doing so must have played a role in their adoption in some cases (although abacus computation is quite rapid and error-free also, as we have seen). So, too, the novelty of the system may have appealed to a newly literate middle class interested in breaking away from older traditions. However, if either of these were the primary explanation for the adoption of Western numerals, we would expect a very different pattern of their retention. We also would not expect them, five hundred years later, to be nearly ubiquitous worldwide, unless they were actually the best system imaginable. That sort of teleological argument may be comforting to someone who prefers to think that we live in the best of all possible worlds, but there is no evidence, other than our own wishfulness, to suggest that it is true.

The most parsimonious explanation for the rise of the Western numerals to their present level of ubiquity is the combination of new technology (the printing press), new forms of information transmission (higher literacy, educational texts), and a set of social and political changes to go with it. At or around the same time as the Western numerals were adopted, the centrality of the Western European countries at the core of a system of domination, communication, and information was coming into place. This new network—the capitalist world system—starting around AD 1450–1600 established itself not just as one of many local networks, but as the principal vector through which information was transmitted and inequality sustained (Wallerstein 1974). Once this network had been established, communication systems whose utility was evaluated within that context became increasingly common and, thereby, increasingly difficult to replace. This is, again, very similar to the argument that Saxe (2012) raises about the transformations and importance of the Oksapmin body-counting practices discussed earlier. It is not the technique or the technology alone, but only taken in conjunction with social, economic, educational, and political systems, that makes the difference.

It is not a coincidence that the Western numerals came to predominate at the same time as the world system. There is no cognitive revolution associated with decimal place value notation. Rather, the confluence of economic, social, and communicative factors cemented both the Western numerals' place as a notation and the economic system that would contribute to its spread. Their ubiquity, established by printers disseminating notations to a new group of literate users, was ensured. Only in such novel circumstances could a break from the Roman numeral tradition be permanently achieved.

Conclusion

In their treatment of frequency dependence, Efferson et al. (2008) show that even in experimental contexts where frequency dependence, particularly conformity, might be expected, many individuals are neither conformists nor passive nonconformists but mavericks. This is a valuable insight because it recognizes that the value of conformity for an individual rests on the presumption that some individuals do not conform to the majority. The authors are also very careful to distinguish conformity, as a disproportionate tendency to follow the majority, from frequency dependence in general. Yet it rests its theoretical focus on outcomes rather than the processes underlying those outcomes, and as such, like other recent work in the tradition of evolutionary studies of cultural transmission (e.g., Rendell et al. 2010), does not recognize important variability in the concept of frequency dependence. The very best work in this tradition, such as Mercier and Morin's (2019) survey of experimental research on conformity and "majority rules" principles, highlights a real challenge to individual adopters, which is that communication is not always to be trusted and so individuals don't always have a good guide as to how to proceed when considering some popular innovation. Frankly, most psychological studies of frequency dependence, being reliant on experimental, short-term conditions with research participants who are not invested in conforming to the majority for any reason—and surely not for networking purposes—are unlikely to shed light on situations like the decline of the Roman numerals and their replacement. I do not see any way to get undergraduates in a laboratory to think like a sixteenth-century printer, much less to recapitulate a centuries-long process of replacement. These experiments will provide insights and hypotheses that will be extraordinarily useful, but that can only be tested through the ethnographic and historical records.

Stephen Shennan (2002: 48) notes that evolutionary anthropologists' and archaeologists' knowledge of the psychological mechanisms behind cultural transmission is rudimentary. He suggests instead that "the way forward for archaeologists and anthropologists, if not for psychologists, seems to be to ignore the psychological mechanisms and accept that, whatever they may be, they lead to culture having the characteristics of an inheritance system with adaptive consequences." Because prehistoric data are rarely amenable to "paleopsychology," it is tempting to fall back on a position that neglects intentionality, or to assume that cultural differences will make it impossible to discern intentionality. If frequency-dependent biases are relatively common, however, then this assumption is not warranted. Many scholars of an earlier generation—sociologists such as Everett Rogers (1962), geographers such as Torsten Hagerstrand (1967), psychologists such as Solomon Asch (1955), and anthropologists such as Margaret Hodgen (1974) and Homer Barnett (1953)—used ethnographic, social, and historical data to conceptualize cultural transmission as the outcome of decision-making processes. This same sort of holistic approach, which does not hide from modeling, evolutionary thinking, or quantitative data but recognizes their limitations, is needed to understand the cognitive underpinnings of historical phenomena like these.

Linguists and scholars of communication have spent less energy than is needed on questions of motivation and decision making in the transmission of variants. Thus, the study of transmission may begin with a return to Zipfian principles—focusing on the adoption and retention of already-common forms—but must quickly move past it. Zipf's analysis does not focus chiefly on the choice between two or more variants, but simply on the description of rank-ordered frequencies. But if we want to understand the choice between *tsunami* and *tidal wave*, *sneaked* and *snuck*, *dozen* and *twelve*, we must investigate how new variants emerge and are accepted in a population, bearing in mind that this is both a cultural and a psychological process as well as a linguistic one. We can also use the models described above to investigate choices between entire communicative systems, be they alphabets, computer operating systems, social networking websites, or numerical systems. Because the adoption of common communication systems becomes more useful as systems acquire more adopters, the communicative sciences constitute a special domain of inquiry relevant to cultural analyses of frequency-dependent phenomena.

The integrative and rigorous understanding of cultural transmission requires that we begin to move away from models that assume that we can know nothing about variation in decision-making processes. The three types of frequency-dependent bias do not merely reflect three different patterns of trait transmission, adoption, retention, and abandonment; they reflect the different decision-making processes described in the three cases above, and potentially in others. These processes remain undertheorized and incompletely demonstrated, but without asking these questions we are unlikely to get an adequate picture of the distinctions among these three types of frequency dependence, and potentially all sorts of other cultural evolutionary processes. We can know more about decision making than we are presently willing to admit, and we have the data at hand to ask useful questions about decision making that will aid in the understanding of evolutionary processes in cultural transmission.

Frequency-dependent biases, under conditions of massively increasing literacy rates and new technologies such as the printing press, structured, if not determined, the replacement of the Roman numerals by the Western numerals in Western Europe. Thus, simply comparing the structure of the Western numerals to the Roman numerals is of limited use, except in narrow contexts such as arithmetic textbooks. While this is an important subject in the history of mathematics, it is not nearly as important in the history of number systems, because most number systems were simply never used for arithmetic, and even those that are, like the Western numerals, are mostly used for representation, not computation. This should also imply that elsewhere in the world, the replacement of nonpositional systems by the Western numerals, or by other decimal, ciphered-positional systems like the Arabic, Indian, Lao, or Tibetan, will have a different trajectory. From a narrowly Western-centered point of view, it's easy to forget that these other systems existed: alphabetic systems such as the Armenian,

Georgian, Arabic, Hebrew, and Syriac numerals; alphasyllabic systems such as the *katapayadi* numerals of the Indian tradition; additive numerals of the southern Indian subcontinent such as the Tamil, Malayalam, and Sinhalese systems; the classical and other traditional numeral systems of China and East Asia. How were these replaced?

If the structure of systems were the most pertinent factor, then decimal ciphered-positional numerals should be equally potent as agents replacing nonpositional systems worldwide, but this is not the case. In China, mathematicians became aware of ciphered-positional numerals in the eighth century CE; Qutan Xida, a Buddhist astronomer of Indian descent working at the Tang Dynasty capital of Changan, wrote his *Kaiyuan zhanjing* between 718 and 729 and described positional numeration with the zero (Needham 1959: 12). But zero wasn't used at all in Chinese arithmetic until the *Shùshū Jiǔzhāng* (Mathematical Treatise in Nine Sections) of 1247, half a millennium later (Libbrecht 1973: 69). Even then, zero didn't serve as a general-purpose number in such texts, but simply to fill in empty medial positions in numbers like 105 or 12,007. The contemporary Chinese zero, *ling* (零), came into use in the late sixteenth century, sporadically at first, but the contemporary Chinese practice of using the traditional numerals 1–9 along with either *ling* or a circle did not become customary until the twentieth century, by which time Western mathematical practices and global mathematical transmission were predominant. As we saw with the Roman numerals, the fact that some innovation is known and might have some advantage does not automatically recommend it to new users.

In India, the homeland of positional numeration, where decimal ciphered-positional numerals were used by the sixth century and perhaps even earlier, nonpositional systems were retained, not just incidentally or for peripheral or prestige purposes but as part of core representational practices. In northern India the transition to ciphered-positional numeration was fairly rapid, with the older Brahmi ciphered-additive/multiplicative-additive system being replaced by the ninth or tenth century by simply retaining the old signs for 1 through 9 and adding a zero. But in southern India, although positional numeration was known very early, it did not replace older notations. The Tamil, Malayalam, and Sinhalese core numeral systems were all ciphered-additive or multiplicative-additive through the nineteenth century; the philologist Pihan (1860) reports all of these systems in use at the time he was writing. Babu (2007) provides an important historical analysis of the teaching of Tamil mathematics, including these numbers, in eighteenth- and nineteenth-century schools. These systems were only replaced in the nineteenth and early twentieth centuries, under British colonial rule (they are retained and used historically, like Roman numerals). A thousand years of contact with northern India did not displace them; a hundred years of colonialism, mass literacy, printing, and education were necessary.

As we saw in the previous chapter, the Arabic world was similarly divided; positional numeration was known there by the seventh century CE, but the ciphered-additive alphabetic numerals, finger reckoning, and other techniques survived far longer. The Arabic abjad numerals are like the Greek alphabetic numerals and other similar systems, in which the letters, in their order, are given the values 1–9, 10–90, and 100–900, with some variation. It seems that, following the advice of al-Uqlidisi which we saw in chapter 3, some Arabic writers may have been doing calculations on a dust board positionally, but permanent records were not common in that system until much later than might be expected. Until the thirteenth century, Arabic astrology and astronomy was conducted using the abjad numerals alone, and for centuries thereafter abjad numerals and positional numerals were used side by side (Lemay 1982: 385–386). The retention of abjad numeration for astronomical and astrological purposes, in which arithmetic is needed intensively, belies any easy functional explanation. Only around the seventeenth century, and especially after the French conquest of Egypt in 1798 and the beginnings of the Arabic printing tradition with movable type, did the abjad numerals come to occupy an auxiliary role similar to the Roman numerals in modern Europe and North America.

In each of these regions, positional numeration was known centuries before it was known in Western Europe, but in all of them, additive systems persisted, not just for a century or two but for a millennium or more in active, regular use by discourse communities. In these communities, the costs of switching from one system to another—the time to learn a new system, the loss of the ability to read older material, and the inaccessibility of new notations to old audiences—overwhelmed any advantage that might have accrued from the switch. Just as with the Roman numerals, it took a disruptive event in the history of those communities to lead to a permanent abandonment of additive systems. The advent of widespread literacy and the printing press, and the integration of local economies and social institutions into global systems, played that role. None of the older systems was forcibly extirpated by colonial officials or imperialist bureaucrats (in contrast to Mesoamerica and the Andes, where Aztec, Maya, and Inka practices were sharply criticized and forcibly curtailed). But the logic of worldwide networked economic and social systems does not always require such direct violence in order to disrupt longstanding systems. It didn't need it in Western Europe, the very heart of the contemporary global world system, to displace the Roman numerals, so why should it have anywhere else?

The one remaining question is: What about new innovations? In figure 4.3, I hypothesized that once a networked frequency-dependent bias takes hold, new innovations that might compete with the dominant system will be developed, but normally quickly abandoned. In the following chapter, we will examine one such case study in detail, in the broader context of the development of numerical notation since the Western numerals achieved their present level of ubiquity.

5 / V Number crunching

The decline in the use of the Roman numerals from 1300 to 1800, accelerating rapidly starting in the seventeenth century, is a complex story occurring across multiple continents and multiple contexts at varying rates. But not all numerical notations fall from such great heights of popularity. More often, when a notation is rejected or abandoned, the process is more rapid. Systems may come into existence within only one or a few communities, among a limited number of writers, only to fade within a generation. Because frequency dependence plays such an outsized role in the adoption and survival of communication systems, new notations—those perhaps developed by a single user or a small group, used locally—are most vulnerable to being snuffed out by more widespread, more generalized systems. Most of these systems are almost invisible at archaeological and longer historical time scales, due to the vagaries of survival and recovery of these kinds of materials. The farther back in time we go, the less likely we are to have reliable evidence for them, because they were never widely used and the evidence for their use survives less and less well. The Roman numerals, then, are the most audacious, and thus the least typical, example of a numeral system going out of regular use.

This chapter looks at the Cherokee numerals, a numerical notation system that never came into common use, and which has actually been so sparsely attested that it is unmentioned in most of the systematic histories of writing and number systems. Cherokee (Tsalagi Gawonihisdi) is a southern Iroquoian language, spoken by around 12,000 people today mainly in Oklahoma and North Carolina, out of a nation of over 300,000 tribal citizens. The history of the Cherokee nation is centuries-long and extraordinarily complex, and inevitably became intertwined with the history of American expansion and aggression in the southern and Plains states

(Perdue and Green 2007). The Cherokee language is distantly related to the northern Iroquoian languages such as Mohawk, Seneca, Oneida, and Cayuga spoken around the eastern Great Lakes and in the St. Lawrence River basin in southern Canada. The political and social history of the modern Cherokee is linked to events farther south, most notably the policies of Andrew Jackson's administration bringing about the forcible relocation of the Cherokee and several other powerful southeastern American nations in the 1830s to what was then known as the Indian Territory west of the Mississippi, and today known as the state of Oklahoma.

The Cherokee writing system, one of the only indigenously developed writing systems of the United States, enjoys a certain pride of place in the history of writing systems and literacy. Starting around 1810 and working over the next decade, the silversmith and intellectual Sequoyah, or George Guest (ca. 1770–1843), developed a writing system of 85 syllabic characters for representing the Cherokee language (figure 5.1). Purportedly, Sequoyah did so without any knowledge of English writing other than (it is sometimes said) a vague awareness of printed newspapers. The debate over how much English he knew, and how much contact he had had with the literate traditions of early American life, has been a heated one in writing systems studies over the past decade (Cushman 2012).

These debates are important in their own context, but sometimes seem to miss the point. We do not impose on other script inventors the requirement that they be completely nonliterate to praise their inventions. We do not say of King Sejong of Korea (1397–1450), for instance, that his fifteenth-century invention of the *hangul* writing system (now the most common writing system in Korea), a highly phonetic writing system, is any less impressive because its inventors and promulgators were previously literate. Attributions of nonliteracy to indigenous peoples, especially First Nations peoples of the Americas, often serve as a romanticized index of pristineness. Anything that smacked of impurity or borrowing was seen, in nineteenth- and much of twentieth-century anthropology, history, and archaeology, as being unworthy of interest, given the romanticized views of indigenous peoples as static (Trigger 1980). Whereas for Western thinkers, standing on the shoulders of giants (as Newton acknowledged himself to do) is a humble but empowering narrative of technological progress, Native American intellectuals, in particular, were and still sometimes are perversely regarded through a racist intellectual framework as improper inheritors of Western

Figure 5.1
Sequoyah (ca. 1770–1843) with his syllabary of 85 characters

knowledge, not really entitled to employ it as a source of innovation. You can see this in its most overt form when Anglo-Americans in North America insist that to be entitled to land and resource rights, First Nations ought to abandon Western-developed technologies—as if somehow being descended from European settlers entitles one specially to the fruits of one's imagined ancestors' innovation. What is innovative about Sequoyah's development of the syllabary is not what he knew or did not know prior to developing it, but that it is structurally distinct from all these potential influences—no other contemporary script he could plausibly have known is syllabic. We can then apply the same principle to his less well-known, and dramatically less widespread, invention of Cherokee numerals.

The Cherokee case shows that the same sorts of factors apply to the selection of new systems as to the decline of longstanding and widespread systems like the Roman numerals. But this is not a narrative merely of decline. Rather, I will show that the same sorts of comparative and cognitive approaches can be applied fruitfully to what is, as it turns out, our best evidence for the development of new numerical notations. With a little linguistic ingenuity, a little knowledge of the cognitive underpinnings of number symbols, and a little awareness of the relationship between graphic systems and systems of power and inequality cross-culturally, we can see the Cherokee development not merely as an abortive attempt quickly abandoned, but as evidence for the ongoing capacity and tendency for humans to innovate in this domain.

Sequoyah's numerals

By 1828, Sequoyah was at the peak of his renown. His syllabary had been accepted by the Cherokee National Council at its national capital in New Echota, Georgia, in 1825. The laws of the Cherokee nation were printed using the syllabary in 1826, while the bilingual, biscriptal newspaper the *Cherokee Phoenix* began printing in 1828 in New Echota using a newly designed typeface. In that same year, Sequoyah traveled to Washington as part of a delegation to advocate against the removal of those Cherokee still living east of the Mississippi to the Indian Territory (now eastern Oklahoma), an effort that would ultimately prove unsuccessful in light of President Andrew Jackson's policies against Native Americans. While there, he met with the literary scholar Samuel Lorenzo Knapp, with whom he communicated through interpreters. Based on this brief and mediated encounter, Knapp immediately perceived Sequoyah's genius and wrote extensively of him as an American philosopher, as part of his *Lectures on American Literature* that appeared the next year (Knapp 1829).

By this time, Sequoyah was becoming widely recognized by the American intelligentsia for his invention of the Cherokee script, the first and at the time only indigenously invented script for any people north of Mexico, but also as a public, *American* intellectual whose accomplishments demonstrated that Europe had no monopoly on inventive genius. This counter-narrative to the myth of the indigenous primitive was never predominant, but at times, when Euro-Americans saw their interests in terms of distinguishing themselves from Europe, some might coopt Native American accomplishments by claiming them as generically American. While, naturally, Knapp's account focused heavily on the syllabary, in this previously unremarked-upon passage he also commented on Sequoyah's arithmetical genius, mentioning for the first time a set of numerical signs Sequoyah is said to have invented to accompany his script:

> He did not stop here, but carried his discoveries to numbers. He of course knew nothing of the Arabic digits, nor of the power of Roman letters in the science. The Cherokees had mental numerals to one hundred, and had words for all numbers up to that, but they had no signs or characters to assist them in enumerating, adding, subtracting, multiplying or dividing. He reflected upon them until he had created their elementary principle in his mind, but he was at first obliged to make words to express his meaning, and then signs to explain it. By this process he soon had a clear conception of numbers up to a million. His great difficulty was at the threshold, to fix the powers of his signs according to their places. (Knapp 1829: 28)

Shortly after his return from Washington—perhaps in late 1829 or early 1830—Sequoyah presented this next invention to the Cherokee National Council: a set of numeral signs (figure 5.2). Unlike the Western numerals in common use, Sequoyah's numerals had principally what I have called in chapter 1 a *ciphered-additive* structure (Chrisomalis 2010: 344). That is, instead of place value and a zero, there are separate signs for each decade and unit, which combine together, so that 67 would be the sign for 60 followed by 7, rather than 6 followed by 7 as in Western numerals. Beyond 100, the system became *multiplicative-additive*—instead of developing nine new signs for 100 through 900, Sequoyah invented only one, which combined with the signs for 1 through 19. Ciphered-additive notation, though rare today, was extraordinarily widespread cross-culturally prior to the past

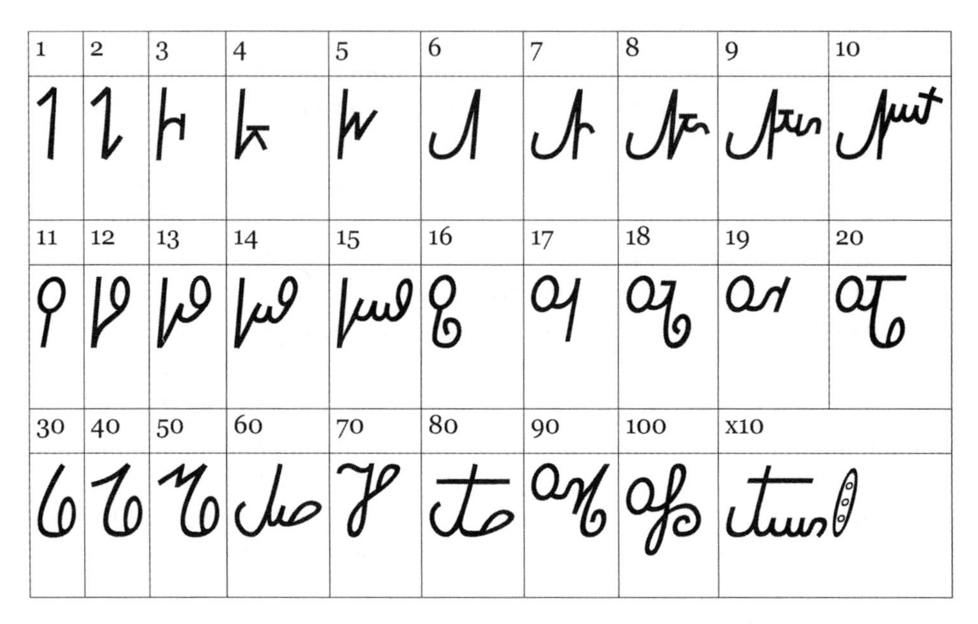

Figure 5.2
Cherokee numerical notation developed by Sequoyah

couple of centuries. Greek, Hebrew, and Arabic alphabetic numerals are all ciphered-additive; all of the mathematics of the ancient Greeks was written down using ciphered-additive numerals, showing that they are perfectly adequate for such functions. From the Egyptian hieratic numerals used in almost all the quotidian tasks of the Egyptian state, to the traditional Sinhalese numerals of south India and Sri Lanka, or the Siniform numerals developed for the Jurchin script in twelfth-century China, ciphered-additive numeration is cross-culturally recurrent. We need not take Knapp's statement that Sequoyah knew nothing of Roman or Western numerals as true simply because he says so, but these other systems were almost entirely unknown in American intellectual life at the time. In other words, just as Sequoyah's syllabary had no parallel among the American scripts of the time, his numerals were similarly distinct from any other local system in use.

Several additional properties of this system demand our attention. Firstly, the numerals 11 through 19 have distinct signs, rather than being combinations of 10 plus one of the signs for the units. Secondly, the first twenty numeral signs are clearly subdivided visually into groups of five (1–5, 6–10, 11–15, 16–20), with a basic graphic unit shared among each group of five, ligatured to a unique element for each separate numeral. Finally, there is a sign that indicates that the value of the preceding sign is to be multiplied by ten. While the Western numerals have no such feature, almost all of the alphabetic systems just mentioned have multiplier signs. They're useful because, in a system without a zero, multiplying using one extra sign is easier than developing an entirely new set of nine signs for the next power of the base. So, for instance, in Greek alphabetic numerals, a subscript stroke or *hasta* placed before a numeral indicated that its value should be multiplied by 1,000—compare β for 2 with ͵β for 2,000 (Threatte 1980).

Today the Cherokee numerals survive principally in a manuscript held at the Gilcrease Museum in Tulsa, Oklahoma, which Sequoyah wrote in late 1839 and one page of which (figure 5.3) was annotated in the English language and Western numerals by his interlocutor, the poet and dramatist John Howard Payne.

By this point, although the Cherokee had been displaced to Oklahoma in the American imperial project now known as the "Trail of Tears," the syllabary itself was widely accepted—by many accounts, the Cherokee literacy rate in the syllabary was as high as Anglo-Americans' was in English at the time. Cherokee writers engaged in a wide variety of formal and vernacular literacy practices, such as the wall inscriptions in Manitou Cave, Alabama, interpreted as relating to ceremonial activities at a time of deep cultural disruption (Carroll et al. 2019). Yet, while Sequoyah remained a figure of prominence, his numerals were rejected by the Cherokee themselves, and all these early texts employ Western numerals when numerals are needed. Other than in John Howard Payne's papers containing the numerals, there is no further nineteenth-century direct evidence for their use, and scant evidence for their origins. Were they rejected because, as Holmes and Smith (1977: 293) argue in their instructional grammar of Cherokee, "They sensibly voted Sequoyah's numbers out, as Arabic numerals, which are simpler, were already in use"? Other than an occasional sentence in one of the many books on Cherokee writing or in biographies of Sequoyah himself, Sequoyah's numerals are largely absent from the narrative of the syllabary and its adoption. At best, they are considered a brief and abortive attempt wisely avoided, the fruit of a "second novel syndrome" in the inventor of graphic notations.

Although the period from 1820 to 1840 is hardly the distant past, we are thus faced with a lack of evidence and a wealth of speculation on several

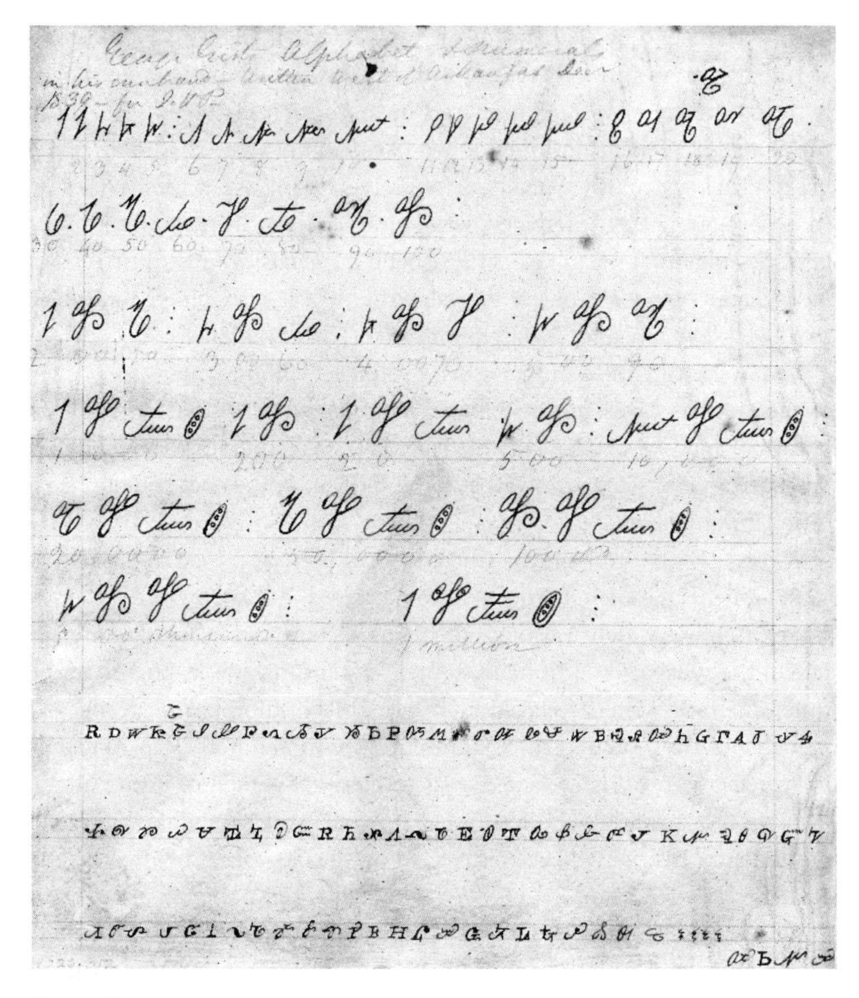

Figure 5.3
Sequoyah's numerals as annotated in 1839 by John Howard Payne (courtesy of Gilcrease Museum, Tulsa, OK, accession number 4026.312)

critical issues. We do not know the answers to five central questions relating to Sequoyah's numerals. First, why do Sequoyah's numerical graphemes below 20 group into fives? Second, why do 11 through 19 have distinct graphemes even though there is no need for them once you have signs for 10 and the units one through nine? Third, what impetus exists for developing a ciphered-additive system where the signs for 20 through 90 have their own distinct signs, and where there is a multiplier for 10 in place of a zero? Fourth, given the success of Sequoyah's syllabary, why were his numerals not similarly successful—why were they rejected by the Cherokee themselves? Finally, why were they invented at all—what are the historical and psychological contexts under which such inspired acts of invention occur? The first three of these questions are structural, based on the lexical and graphic properties of language and notational systems, while the latter two are social and cultural.

One obvious source for any structure of a numerical notation is the lexical numeral system of its inventors and its users. Perhaps some of the things that we think irregular about Sequoyah's numerals are understandable in light the morphology of the Cherokee number words. Table 5.1 has a modern linguistic rendition of the Cherokee numerals, which differs only in slight details from those presented in nineteenth-century grammars (Montgomery-Anderson 2015). We can see that with only one partial exception, the Cherokee verbal numerals are an ordinary decimal number system in which the powers of 10 (*sgoóhi* for ten and *sgohitsgwa* for hundred) are combined with the ordinary words for one through nine. The only irregularity is a common one cross-linguistically, which is that the teens from 11 through nineteen are not simply "ten one, ten two, ten three," but use a special morpheme, *du,* which seems to mean "10" but is perhaps better translated as "beyond" or "added on," with the ten elided. While these are transparent, they are irregular, just as the teens are in most Indo-European languages. This does suggest that this feature, at least, was salient to Sequoyah when inventing his numerals.

Most numerical notations do not reproduce the irregularities of their speakers' number words. So, for instance, English *eleven* and *twelve* are opaque (they are ultimately etymologically "one left" and "two left"), but

Table 5.1
Cherokee numeral words (after Montgomery-Anderson 2015)

1	saàgwu	11	sáʔdu	10	sgoóhi	100	sgohitsgwa
2	táʔli	12	taldu	20	talsgoóhi	200	talisgohitsgwa
3	joʔi	13	joogádu	30	joʔsgoóhi	300	joʔsgohitsgwa
4	nvhgi	14	nvhgádu	40	nvksgoóhi	400	nvksgohitsgwa
5	hisgi	15	hisgádu	50	hiksgoóhi	500	hiksgohitsgwa
6	suúdáli	16	suudaldu	60	suudalsgoóhi	600	suudalsgohitsgwa
7	gahlgwoógi	17	gahlgwoodu	70	gahlgwasgoóhi	700	gahlgwasgohitsgwa
8	chaneéla	18	chanelaadu	80	nelsgoóhi	800	chaneélasgohitsgwa
9	sohneéla	19	sohnelaadu	90	sonehlsgoóhi	900	sonehlsgohitsgwa
10	sgoóhi	20	talsgoóhi	100	sgohitsgwa	1,000	iyágayvvli

the numerals 11 and 12 follow the ordinary base-10 Western numeral system. But there are definitely some other cases where lexical irregularities are reproduced in a numerical notation, and the teens figure among those. In ciphered-additive Greek alphabetic numerals for 11–19, we would expect the sign for 10 (iota, I) to be followed by the signs for 1–9, because, as discussed in chapter 1, most numerical notations are written in descending order, starting with the highest powers. But, instead of writing 12 as Iβ, 10 + 2, some ancient Greek texts would write it βI, 2 + 10, to correspond with the numeral word *dodeka* "two (plus) ten" (Threatte 1980).

The Jurchin numerals, mentioned above, were designed for use alongside the script of the Jurchins, speakers of a Tungusic language in northeastern China. The Jurchins ruled much of northeastern China from around 1115 to 1234, right up to the Mongol invasions, and while their script and numerals draw from Chinese influences (the script is called "Siniform" because of its visual and structural similarity), its numerals are quite distinct from the classical Chinese numerals laid out in figure 1.1. The Jurchin numerals are ciphered-additive but also have special signs for 11 through 19, exactly paralleling the structure of Cherokee (figure 5.4). In both the Cherokee and Jurchin cases, the reason for 11–19 receiving distinct signs derives from the words themselves. The Cherokee word *sgoóhi* for 10 is not a morpheme found in any of the words for 11–19, which instead use *-(á)du*—which Montgomery-Anderson (2015: 178) simply describes as an "additional element" but is definitely not a form meaning "ten." We might think that a

1	2	3	4	5	6	7	8	9	10
[illegible]	[illegible]	[illegible]	[illegible]	[illegible]	[illegible]	[illegible]	[illegible]	[illegible]	[illegible]
emu	ǰuwe	ilan	duwin	šunǰa	ningu	nadan	ǰakun	uyun	ǰuwa

11	12	13	14	15	16	17	18	19	20
[illegible]	[illegible]	[illegible]	[illegible]	[illegible]	[illegible]	[illegible]	[illegible]	[illegible]	[illegible]
amšo	ǰirhon	gorhon	durhon	tobohon	nilhun	darhon	niyuhun	oniyohon	orin

30	40	50	60	70	80	90	100	1000	10000
[illegible]	[illegible]	[illegible]	[illegible]	[illegible]	[illegible]	[illegible]	[illegible]	[illegible]	[illegible]
gušin	tehi	susai	ninjhu	nadanju	jhakunjhu	uyunju	tangu	mingan	tuman

Figure 5.4
Jurchin numerals

system that has distinct signs for 11–19 is likely to have a base of 20, but that was not true of Cherokee—it was, rather, that the words for the teens were special, within an otherwise generally decimal verbal numeral system.

One motivation for the development of ciphered-additive notation may be morphological. In many decimally structured languages, the words for the tens are more opaquely built from the units plus the morpheme for "ten"—so, for instance, Latin *viginti* is not self-evidently "two tens." The Cherokee decades, however, are not opaque. Could there instead be a cognitive reason for using the ciphered-additive structure? Drawing from the insights of chapter 2 on Zipf's principle of least effort, ciphered-additive systems are more concise than a place value system like our own for writing any number with a zero in it. Where we need two signs to write 40, Sequoyah needed only one. There is a direct connection between conciseness and frequency in ciphered-additive systems, just as there is in cumulative-additive ones like the Roman numerals. In decimal languages, numbers with zeroes, the round numbers, are used more frequently (so for instance 100 is more often used than 99 or 101 in almost any language with a decimal structure). On the other hand, the cognitive cost is twofold. First, the semantic and arithmetical relationship between 4 and 40, 7 and 70, and so on, is lost in a ciphered-additive system—these pairs of signs bear no visible relationship to one another. Second, one needs many more signs to make the system work: nine signs for 1–9, nine more for 10–90, and so on. For this reason, most ciphered-additive systems switch for higher powers to multiplicative structuring, and here we see again that Sequoyah's system fits right within the cross-cultural norm—it is multiplicative above 100, again like Jurchin, which is perfectly identical to it structurally. That two essentially identical structures would be developed, centuries apart, with absolutely no cultural contact between them, and that both would be highly distinct from the local systems on which they might have been based, is striking. It both confirms the role of constraint in limiting the possibilities for numerical innovation, while at the same time demonstrating that innovation, rather than simple emulation, is expectable and normal in the history of numerical notations.

A quinary quandary

We are still left with the puzzle of why the signs for 1 through 20 are grouped visually into groups of five. In my previous research, I would have attributed this to Sequoyah's clear aesthetic sense—he was a silversmith by trade—and perhaps I would have mused idly about the role of the hands in promoting base-5 or quinary notation.

Looking to the number words, a glance at table 5.1 shows that there was absolutely no quinary, or base-5, component to the Cherokee lexical system. Floyd Lounsbury (1946) reports on what he calls a "stray" numeral system that some Cherokee informants in the early 1940s reported to him, but this consisted of a bare set of nine numeral words that could not even be assigned firm meanings, and in any case it does not have quinary elements. Quinary number word systems are relatively uncommon on their own, but are quite common in combination with vigesimal (base-20) systems (Nykl 1926). Quinary-vigesimal structures are common both in Mesoamerican languages and in the languages of California, but east of the Rockies only a handful of languages, the Caddoan languages such as Pawnee and Caddo, have such a structure (Eells 1913; Nykl 1926). While Caddoan languages were spoken in Oklahoma and parts of the Great Plains, and Sequoyah surely would have known and encountered speakers of some of these languages, there is no reason whatsoever to believe that he spoke to any of them or was influenced by them.

In any case, the Cherokee number words are not really base-20, just as they are not base-5, so this is an unlikely route by which this structure would have entered into Sequoyah's mind. In fact, it's relatively common for decimal languages with ordinary decimal number words to add an element of a subbase of 5 into their numerical notation. For instance, compare the Latin number words *unus, duo, tres, quattuor, quinque, sex, septem, octo, novem, decem* with the Roman numerals I, II, III, IV, V, VI, VII, VIII, IX, X. Latin number words were not the source of V = 5, nor were the number words of Etruscan or Greek or any of the other local languages of fifth-century BCE Italy. Rather, once again we encounter the power of the subitizing limit, discussed earlier, which provides a constraint influencing systems in this direction to avoid needing to write five or more identical signs. But here we run into an obstacle in the Cherokee case: the Cherokee system is ciphered, and so it doesn't rely on the repetition of signs in the way a cumulative system would. So how can cognitive factors affect the structure of its signs?

Luckily, there is some historical evidence giving us some insight into Sequoyah's inventive process that bears directly on this question. In 1835,

Sequoyah's cousin George Lowery, a bilingual English and Cherokee speaker who had been a major during the War of 1812, wrote a biography of Sequoyah, with an introduction and transcription by John Howard Payne (in whose papers the numerals survive), in which he discussed his kinsman's life and practices from a culturally near and personally familiar perspective. In this passage, which has been reprinted twice since its initial publication but never analyzed, Lowery shows us the thought process of an inventor solving the numerical problems of recordkeeping in a series of steps: starting with marks for each different coin, then using tally marks alone, then tally marks crossed after every fifth number, then a special sign X for 10, then finally the fully developed numeral system we have been discussing:

> But he had to give much credit; and could not remember the persons and the sums, there were so many. His alphabet had not then been thought of. He devised this mode of keeping his accounts: He would express the person by some sign or by something like a nude profile or full length likeness: and after it he would put as many large round marks as there were dollars & for smaller sums, as many smaller circles, thus: for $1.75 he would put OOo; but this soon confused him & then he would say for several dollars, five for instance O|||||, and so on; and after that, he would make straight marks, crossing them at every 5 as this ǂǂǂǂǂ, & if there were more tens than one, expressing each ten by a cross X. At last, he made the numbers, as he now uses them. (Lowery 1977)

Bearing in mind that we really have only secondhand accounts here, we can start to see an explanation emerge in several stages. The first, Lowery reports, uses circles of different sizes for coins of different denominations—in his description, $1.00, $0.50, and $0.25, in descending order. This is not precisely a numerical system, but a very innovative iconic notation for silver coins of different sizes. The next step in Sequoyah's thinking seems to employ these iconic depictions alongside a sort of tally-marking system, and then eventually leads to what seems to be a cumulative-additive numerical notation system with ones grouped into fives with a cross across each group of five strokes, and X marks for 10. The fact that Sequoyah eventually used X for 10 suggests, but does not prove, that he knew the Roman numerals, which would have been in general if not everyday use in American institutions at the time. Only after this sequence of inventions, each building on aspects of the previous one, did Sequoyah end up with "the numbers, as he now uses them"—in other words, the ciphered-additive numerals described above. Thus, before he invented his numerals, and perhaps even before he had the idea for the syllabary itself,

Sequoyah was thinking and representing matters visually using fives and tens. It is a small step from there to the idea of retaining a quinary visual structure in the system itself, even after he had abandoned the initial principles under which he was reckoning. We have in this passage one of the only accounts, if not from an inventor himself, then from close kin, of the stepwise and methodical invention of a numerical notation system. We are looking at a micro-scale diachronic process of invention, alteration, and replacement of different notations leading to a final ciphered-additive decimal system.

In most contexts where such problems emerge in contemporary societies, the solution is to borrow some technique or notation from others. And there were surely such models available to Sequoyah, in both the Roman numerals and the Western numerals. I do think, however, that there are good reasons why a Cherokee of inventive mind would develop a new system rather than borrowing another directly. I return to the year where I began, 1828, and the second issue of the bilingual newspaper, the *Cherokee Phoenix*, published by Elias Boudinot, a Cherokee intellectual with a strong sense of national identity and whose paper embodied the aspirations of America's indigenous elites. Boudinot starts one column by reprinting a frankly racist column on "Indian Arithmetic" from the *Western Review*, repeating a set of common nineteenth-century Anglo ideas about the mathematical language and capacities of native peoples (figure 5.5). Immediately below this, Boudinot printed without commentary a list of three columns: the Cherokee lexical numeral words as high as one million (written in the syllabary), the Western numerals, and the Roman numerals (which in their modern forms, top out at 1,000) (figure 5.6). And immediately below that, the third and final plate in the column, he reprints a pair of quotations on the theme of revenge, and the power of silence in the face of a slight (figure 5.7).

This is a clever piece of rhetoric, an intertextual narrative aimed directly at the growing bilingual, biscriptal Cherokee educated audience who were Boudinot's readership. The first article sets out the dominant American narrative of the primitivity of indigenous Americans, which would have been sadly familiar to Cherokee readers. The next is a straightforward empirical demonstration of the falsehood of the first, juxtaposing the long sequence of Cherokee numeral words (though not the numeral symbols, which were

Indian Arithmetic.—Their manner of numbering evidences the extreme simplicity of their language. We have asked of all the tribes, with which we have met, their numerical terms, as far as a hundred. In some few, the terms are simple, as far as ten. In others, six is five-one, seven, five-two. and so on. Beyond ten, they universally count by reduplication of the tens. This they perform with great dexterity by a mechanical arithmetic, intricate to explain, but readily apprehended by the eye. The principal operations are bringing the open palms together, and then crossing the hands, which tells as far as a hundred.—Some of the tribes are said to be perplexed in their attemps to number beyond a hundred.—When the question turned upon any profit, that involved great numbers, we have generally heard them avail themselves of an English word, the first, we believe, and the most universally understood by savages—'heap!'—*Western Review.*

Figures 5.5–5.7
Three plates from the second issue of the *Cherokee Phoenix*, 1828

CHEROKEE NUMBERS.

ᏗᎪᏪᏩᏗ ᏗᏎᏍᏗ.

ᏴᎾᏁᎬ ᏧᎾᏎᏍᏗ ᏓᏃᏪᎵᏍᎬ ᏔᎵ ᎢᏳᏓᎴ ᏂᏓ-ᏅᏁᎸᎢ. ᎬᏂᏗᎾ ᏭᎾᏂᏗᎵ ᏂᏕᎬᏅ ᏕᎪᏪᎳ.—ᎤᏚᎵᏗᏉᏍᎩᏂ ᎢᏗᏩᎳᎩ ᎾᏍᎩ ᏗᎦᏕᏩᏆᏍᏗᏱ; ᎠᎯᏗᏰᏃ ᏗᎪᏪᏩᏗᏱ ᎠᎴ ᏗᎦᏕᏩᏆᏍᏗᏱ.

ᏌᏊ	1	I
ᏔᎵ	2	II
ᏦᎢ	3	III
ᏅᎩ	4	IV
ᎯᏍᎩ	5	V
ᏑᏓᎵ	6	VI
ᎦᎷᏉᎩ	7	VII
ᏧᏁᎳ	8	VIII
ᏐᏬᏁᎳ	9	IX
ᎠᏍᎪᎯ	10	X
ᏌᏚ	11	XI
ᏔᎳᏚ	12	XII
ᏦᎦᏚ	13	XIII
ᏂᎦᏚ	14	XIV
ᏍᎩᎦᏚ	15	XV
ᏓᎳᏚ	16	XVI
ᎦᎷᏆᏚ	17	XVII
ᏁᎳᏚ	18	XVIII
ᏐᏬᏁᎳᏚ	19	XIX
ᏔᎳᏍᎪᎯ	20	XX
ᏐᎢᏦᏁ	21	XXI
ᏔᎵᏦᏁ	22	XXII
ᏦᎢᏦᏁ	23	XXIII
ᏦᎠᏍᎪᎯ	30	XXX
ᏦᎠᏍᎪᎯ ᏐᎢᎦᎵ	31	XXXI
ᏦᎠᏍᎪᎯ ᏔᎵᎦᎵ	32	XXXII
ᏅᎦᏍᎪᎯ	40	XL
ᏅᎦᏍᎪᎯ ᏐᎢᎦᎵ	41	XLI
ᎯᏍᎦᏍᎪᎯ	50	L
ᏑᏓᎳᏍᎪᎯ	60	LX
ᎦᎷᏆᏍᎪᎯ	70	LXX
ᏁᎳᏍᎪᎯ	80	LXXX
ᏐᏬᏁᎳᏍᎪᎯ	90	XC
ᎠᏍᎪᎯᏧᏈ	100	C
ᏔᎵᏧᏈ	200	CC
ᏦᎢᏧᏈ	300	CCC
ᏅᎩᏧᏈ	400	CCCC
ᎯᏍᎩᏧᏈ	500	D
ᏑᏓᎵᏧᏏ	600	DC
ᎦᎷᏉᎩᏧᏈ	700	DCC
ᏁᎳᏧᏈ	800	DCCC
ᏐᏬᏁᎳᏧᏈ	900	DCCCC
ᎠᎦᏴᎵ	1,000	M
ᎠᏍᎪᎯ ᎢᏯᎦᏴᎵ	10,000	
ᎠᏍᎪᎯᏧᏈ ᎢᏯᎦᏴᎵ	100,000	
ᎠᎦᏴᎵ ᎢᏯᎦᏴᎵ	1,000,000	

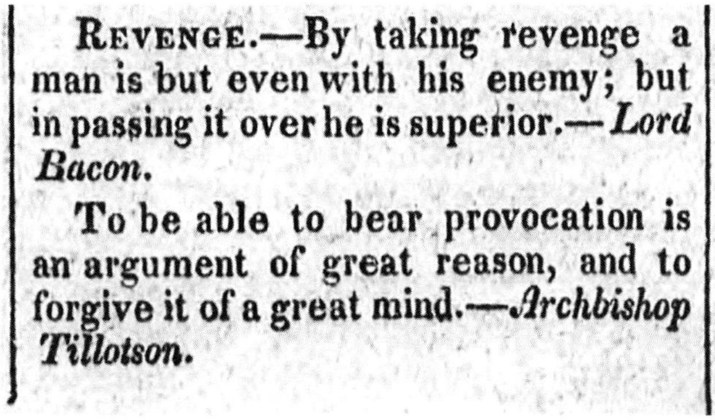

REVENGE.—By taking revenge a man is but even with his enemy; but in passing it over he is superior.—*Lord Bacon.*

To be able to bear provocation is an argument of great reason, and to forgive it of a great mind.—*Archbishop Tillotson.*

Figures 5.5–5.7 (continued)

not part of the typeset for the *Phoenix* and not much in use anyway) with the two dominant notations of the day, the Roman and Western. And, in the last text, a cheeky epigram brings the matter to a satisfying close, in case anyone was confused about this resistance narrative.

We must remember, in all of this, that the Cherokee were (and still are) a colonized people fighting for their claim to an intellectual heritage and to the right to be regarded as civilized. In a context where numeracy indexes civilization, just as literacy does, it should come as no surprise that there is a motivation to develop a local notation rather than borrow an existing one. As we saw in the discussion of conspicuous computation (chapter 2), because numeracy is linked to the power of the state, the symbolic capital attached to numeracy is enormous. In such circumstances, many societies choose to use a borrowed system, and of course that is exactly what the Cherokee National Council decided. Where the paucity of the numerical lexicon is linked to the stereotype of "savage number," as the historian of science Michael Barany (2014) has shown to be widespread in nineteenth-century Western thought, the inventive process is not merely about developing a tool, but developing an emblem that asserts the right to be considered as fully civilized.

Outnumbering the Cherokee numerals

Unfortunately, this context is also the partial answer to why, in the end, the Cherokee chose to adopt Western numerals. The power of the Western numerals is not, as I have argued earlier, that they are structurally superior as a notation. They have certain advantages, but in other respects (such as conciseness), the Cherokee numerals are perfectly adequate and in some cases better than the Western numerals. Rather, the Western numerals' strength in the nineteenth-century American context was their ubiquity, because numeral notations are communication and representation systems first and foremost, *not* arithmetical systems. As we saw in chapter 4, what makes a communication system useful, at a first approximation, is that it is able to be used with many, and particularly many prestigious, individuals. Frequency-dependent biases militate against the adoption of systems used only by a few—even if by someone as locally prestigious as Sequoyah himself. It is rational, in such a context, to adopt a system that is used by many people rather than one used by few, regardless of its other features. Once again we see shades of Zipfian principles. Almost paradoxically, Sequoyah's numerals could not succeed because they had not yet succeeded, and came into existence in a social context where a prestigious, common notation had been adopted almost universally.

But the Cherokee syllabary is also a communication system, and it did survive, if not thrive, throughout the nineteenth and twentieth centuries and to the present, where it is taught and used at least to a limited degree in Oklahoma and North Carolina, where most Cherokee now live (Bender 2002; Cushman 2012). So we need some other factor, other than simple frequency dependence, to explain why the script survived but the numerals did not. Partly the explanation may be biographical. Sequoyah's syllabary had spread quite far by the late 1820s, at which time he began to promulgate his numerals. If the two had been invented and disseminated in tandem, their fates might have been more closely intertwined. Additionally, by the nineteenth century, the Western numerals were extraordinarily widespread and linked to scientific and technical practice. The Roman/Latin alphabet was predominant in American life, but was not presumed to serve as a universal script. For one thing, phonetic scripts record some signs and not others, requiring modification or replacement to serve the speakers of different languages. This is not so with numerical notations, given the virtual universality of the counting numbers across linguistic and cultural divides. But even more important was the value of the Western numerals for entry into the newly developing American educational elite, something that the Cherokee middle class wanted very much. Adopting the syllabary was a nod in the direction of cultural strength and resistance to homogenizing forces. Adopting Sequoyah's numerals, and forgoing the (by this time) universal Western numerals and the arithmetical practices that accompanied them, was a step too far. The Western numerals were regarded, by both colonizer and Cherokee, as part of a prestigious system for economic, technological, and scientific development, really only beginning to flourish in America in the 1830s and 1840s (Cohen 1982).

The failure of Sequoyah's numerals to thrive is not the whole point, however. In short order, he invented not one but two numerical notations, the first cumulative-additive, the second ciphered-additive, neither closely resembling any possible antecedent of which he may have been aware. The invention of the Cherokee numerals was a sequence of inventive steps, built perhaps with some knowledge of other notations used locally, but progressing largely on its own terms. It survived, fortuitously, only in a handwritten manuscript shared between Sequoyah and John Howard Payne, and in the narratives of Knapp and Lowery who reported on the numerals' use at one degree removed from their inventor. We should then ask: how many other numerical notations met a similar fate, with no evidence surviving at all?

Over 5,500 years of written history, there are around 100 numerical notations well-attested enough that I could describe them in my previous work (Chrisomalis 2010), including the Cherokee case. Each of the five basic structures I described earlier was invented multiple times independently of one another. While, in broad strokes, the history of numerical notation converged on one overwhelmingly popular system, the Western numerals, by around 1700 or so, numerical innovation never ceased. What changed was the likelihood that a community would adopt a numerical system widely, because of the overwhelming frequency-dependent and prestige biases in favor of Western numerals.

I conclude with one further comparative case, from thousands of miles away and culturally unrelated. In the middle of the twentieth century, Lako Bodra, a shaman of the Ho people of Bihar province, India, developed a local script and numerical notation called Varang Kshiti (Pinnow 1972) (figure 5.8). The Ho language, of the Munda family, was not previously typically written, although Bodra asserted that Varang Kshiti was not a recent invention but a thirteenth-century script that he rediscovered and modernized (Zide 1996: 616–617). Like the Cherokee, Ho speakers are a minority asserting a cultural identity against long odds. Unlike the Cherokee case, Lako

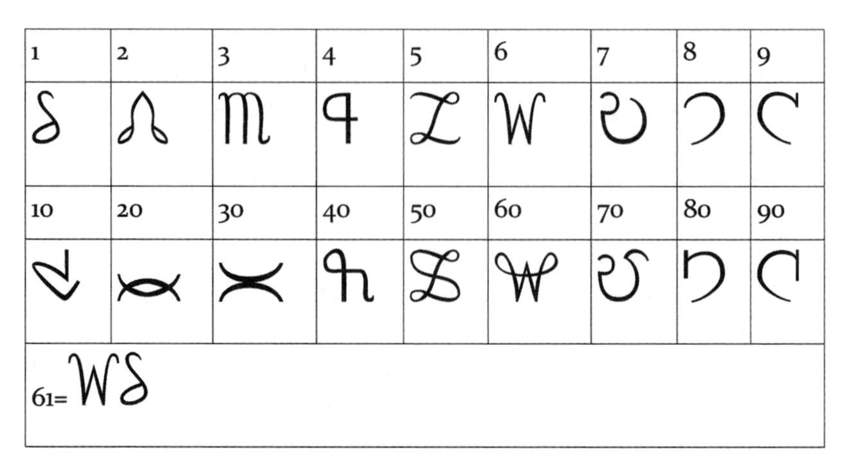

Figure 5.8
Lako Bodra's numerals for Varang Kshiti

Bodra was a literate, college-educated man whose inventions were undertaken while he was working as a railway clerk. As in the Cherokee case, Lako Bodra's script was rapidly adopted by a substantial number of Ho speakers, and it is used in both primary and secondary education. Unlike the Cherokee case, the Varang Kshiti numerals appear to have enjoyed some popularity and retained some use. Strikingly, Varang Kshiti numerals are, like Sequoyah's, also ciphered-additive, with special signs for 10 through 90. While there were, historically, many ciphered-additive numeral systems in the region, none of these survived long enough to be plausible ancestors of Varang Kshiti—unless, of course, its antiquity is as Lako Bodra claimed it to be.

Notations like these confirm the vitality of the inventive spirit motivating individuals and groups in each generation to experiment with numerical forms. They should also alert us to the probability that many, many older systems have been invented and either rapidly abandoned or "outnumbered" by more widespread systems. Few or limited traces may survive of these, and we may not be looking in the right places for them. I would not be surprised if as many numerical notations have yet to be discovered as have been documented to date. Some of these systems may be truly unusual compared to those already known, but the fact that the same structures, like ciphered-additive notation, are invented time and time again suggests that we are not dealing with a situation of extreme variability. Cases like Sequoyah's invention and refinement demonstrate a panhuman interest in numeration, given some stimulus and interest, as well as the cognitive and notational constraints that limit human inventiveness. Most of these innovations will be short-lived, and from a long-term perspective will fail. In contexts like the modern era, where Western numerals are overwhelmingly popular, it may seem pointless to focus our attention on them. But rather than seeing these merely as "dead reckonings," we should instead regard them as evidence of the innovative numerical capacities and interests of our species, and of the brilliant inventors whose intellect underpins them.

6 / VI How to choose a number

In 1792, the wealthy businessman John Jacob Astor wrote one of the first checks issued by the Bank of the United States, newly founded the year before by Alexander Hamilton (figure 6.1). Then as now, the structure of the written text of checks required a repetition of numerals both in numeral words (fifteen hundred and fifty) and in numeral signs (1550), to secure against fraudulent alteration and to reduce the risk of ambiguity of reading. The practice of dual notation is an essential norm of this text genre to this day. We all still do it—those of us who still write paper checks, at least—even if we don't necessarily reflect very much on why. This 200-year-old check thus has a familiarity to the modern eye despite many other differences.

Over a century later, the first edition of the now-famous *Chicago Manual of Style* was published in 1906. In figure 6.2 we see a dizzying array of rules and principles at play to answer the seemingly simple question: How should I write a number? We might prefer conciseness, writing 128 instead of one hundred and twenty-eight. We might prefer aesthetics when starting a sentence with words ("Five hundred and ninety-three"). We are urged to prefer "two dollars" written in words in preparing ordinary reading matter but to use an ideogram (the dollar sign) and numeral sign in matters "of a statistical character," depending on genre. Years should always be written in numerals—even though the reader isn't given any guidance whether to read such a numeral aloud as "two thousand eighteen" or "twenty eighteen" or something else entirely. We might prefer consistency across multiple numerals in a sentence, wherever two or more of these principles conflict.

All published writers have had to deal with the modern versions of these same rules, often to our own dismay, in the editorial process for our publications. Once, when I was working as an editorial assistant for the late Bruce Trigger, my doctoral supervisor, he reported to me with alarm that

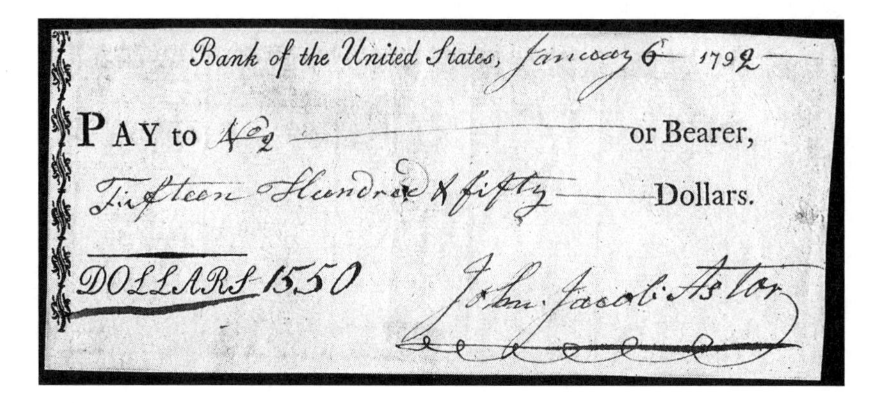

Figure 6.1
Check written by John Jacob Astor, 1792

30 *The University of Chicago Press*

e) Names beginning with "Mc," whether the "Mc" part is written "Mc," "Mac," "M'," or "Mac" without the following letter being capitalized (as in "Macomber"), fall into one alphabetical list, as if spelled "Mac."

84. In ordinary reading-matter, all numbers of less than three digits, unless of a statistical or technical character, or occurring in groups of six or more following each other in close succession:

"There are thirty-eight cities in the United States with a population of 100,000 or over;" "a fifty-yard dash;" "two pounds of sugar;" "Four horses, sixteen cows, seventy-six sheep, and a billy goat constituted the live stock of the farm;" "He spent a total of two years, three months, and seventeen days in jail." But: "He spent 128 days in the hospital;" "a board 20 feet 2 inches long by 1½ feet wide and 1½ inches thick;" "the ratio of 16 to 1;" "In some quarters of Paris, inhabited by wealthy families, the death-rate is 1 to every 65 persons; in others, inhabited by the poor, it is 1 to 15;" "His purchase consisted of 2 pounds of sugar, 20 pounds of flour, 1 pound of coffee, ½ pound of tea, 3 pounds of meat, and 1½ pounds of fish, besides 2 pecks of potatoes and a pint of vinegar."

Treat all numbers in connected groups alike, as far as possible; do not use figures for some and spell out others; if the largest contains three or more digits, use figures for all (see **86**); per cent. should always take figures:

"The force employed during the three months was 87, 93, and 106, respectively;" 1–10 per cent.

85. Round numbers (i. e., approximate figures in even

Manual of Style: Spelling 31

units, the unit being 100 in numbers of less than 1,000, and 1,000 in numbers of more):

"The attendance was estimated at five hundred" (but: "at 550"); "a thesis of about three thousand words" (but: "of about 2,700"); "The population of Chicago is approximately two millions" (but: "1,900,000"). Cases like 1,500, if for some special reason spelled out, should be written "fifteen hundred," not "one thousand five hundred."

86. All numbers, no matter how high, commencing a sentence in ordinary reading-matter:

"Five hundred and ninety-three men, 417 women, and 126 children under eighteen, besides 63 of the crew, went down with the ship."

When this is impracticable, reconstruct the sentence; e. g.:

"The total number of those who went down with the ship was 593 men," etc.

87. Sums of money, when occurring in isolated cases in ordinary reading-matter:

"The admission was two dollars."

When several such numbers occur close together, and in all matter of a statistical character, use figures:

"Admission: men, $2; women, $1; children, 25 cents."

88. Time of day, in ordinary reading-matter:

at four; at half-past two in the afternoon; at seven o'clock.

Statistically, in enumerations, and always in connection with A. M. and P. M., use figures:

at 4:15 P. M. (omit "o'clock" in such connections).

Figure 6.2
Chicago Manual of Style, first edition: rules for writing numbers (University of Chicago Press 1906: 30–31)

the book I was helping him with had been copyedited to replace the word "million" with "10 lakhs" and "two million" with "20 lakhs." It turns out that the editing had been outsourced to an Indian firm, whose copyeditor had duly employed the word "lakh" consistently, which is Indian English for 100,000 (from Hindi *lakh*, hundred thousand).[1]

These sorts of variations, these degrees of freedom in writing and reading texts, are ubiquitous in textual traditions. Over the past chapters, I've written at length about how individuals and groups adopt, use, and abandon particular numerical notation systems in specific contexts. But this perspective presents choice as a binary, as if one day a writer simply decided to stop using Roman numerals and then never used them again. That might conceivably happen, but it isn't the most likely scenario. Skilled writers are always taking account of the context in which, and the audiences for which, they're writing. And because, in addition to numerical notations, languages also have number words available for expressing numbers, the choices and combinations available to writers are even more expansive than simply a choice among notations.

Writing systems typically record language, but numerical notations—like Roman numerals, or like the Western numerals 0 through 9—are graphic, relatively permanent notations not linked to any specific language. We can read them in whatever language we prefer, and often do—in other words, they're not multilingual (recording information in multiple languages) but rather *translinguistic* (allowing the reader to choose what language to read something in). Most writing systems, especially after the earliest ones, can be used to write any word in a language, including the set or sets of lexical number words of the language. Given the ubiquity of lexical and nonlexical resources for expressing numbers in writing, almost any numeral phrase can be written using a variety of strategies.

Here are seven different English expressions for the same number (one million two hundred thousand) from relatively formal texts from the first half of the twentieth century:

> Canadian production of steel ingots from 1935 to 1938 averaged about **1 million, 200 thousand** tons per annum. (Anonymous 1941: 154)

> But perhaps the most important fact bearing upon the situation is the report of the steamship companies to the effect that over **twelve hundred thousand** applications have been received in the last four years for passage to their native European lands, as soon as the war is over. (Freund 1918: 20)

> In fiscal 1948 such carryings under the U.S. flag amounted to 1.9 million tons out of a total of 3.6 million tons, while in 1938 the U.S. flag share was **1.2 million** tons of the 2.4 million tons total industrial carryings. (Anonymous 1949: 82–83)

> There are **one million two hundred thousand** students in high schools; a third of a million in higher institutions. (Cattell 1914: 156–157)

> Whole shrimp examined immediately after removal from the trawl-net varied in bacterial count from 1,600 to **1,200,000** per gram. (Green 1949)

> **One point two million** ($1,198,150), or .6%, for command and management. (Anonymous 1953: 68)

> The total number of ions per cc. produced in air by the radiation to which they are subject is then not more than thirty times $\mathbf{1.2 \times 10^6}$ (the number of seconds in two weeks), or 3.6×10^7. (Muller and Mott-Smith 1930: 279)

We can trace the rise and fall in frequency of different forms for this number over time (figure 6.3). In English, we have the largest and broadest corpus of written texts available for analysis, the Google Books corpus, analyzable through the Ngram tool, which traces changes in the relative frequency of words among 155 billion words in English-language printed books (Google Ngram Viewer 2016, http://books.google.com/ngrams). Throughout the nineteenth century, *twelve hundred thousand,* which to many modern ears sounds jarring or even ungrammatical, was the most frequent form. (I can say *twelve hundred* without an issue, but when I add another number word on the end of it, it sounds infelicitous to me, and to others. This form is almost absent from contemporary written English.) *One million two hundred*

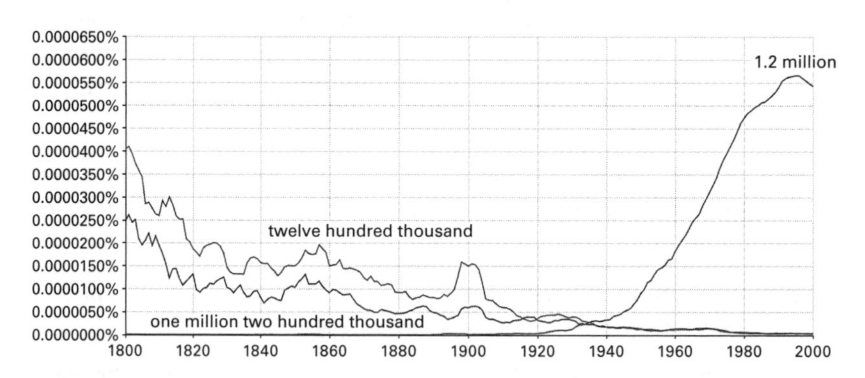

Figure 6.3
Variation in frequency of expressions for "1.2 million," 1800–2000 (Google Books Ngram Viewer, http://books.google.com/ngrams)

thousand was slightly less popular but still widely used. In contrast, *1.2 million* was almost absent from nineteenth-century English books, but especially after 1940 rapidly overtook all other options. Other options, like *one point two million,* are rare throughout the whole period and only show up in the twentieth century. And still other options, like 1.2×10^6, using scientific notation, are vanishingly rare in ordinary books, magazines, and texts intended for nonspecialists.

Some of this linguistic change reflects changing genres—the rise of scientific writing, for instance, affects word frequencies in the Google Books sample (Pechenick et al. 2015). Some of it surely reflects the changes in formal style guides, which influence what makes it into print. And some of it reflects changes in taste and aesthetics, and in what English speakers and writers choose and expect in writing. But at no point has there been just one grammatical way to say or write this number. This variation exists across text genres—for instance, scientific writing may well differ from literature—but also may vary by time period, the nature of the text's production (e.g., transcription or editing), and the individual idiosyncrasies of writers. How do literate norms and traditions, and all these other factors, constrain user choices? We are once again returning to the issues of constraint outlined in chapter 1, but with additional ammunition at our disposal. Recognizing that there is variation is only the first step. Determining how genre, dialect, and broader social factors interact with the preferences of individual writers and readers must follow.

Explaining how and why these sorts of variability persist, and why these changes occur, is no easy task, even for a contemporary world language like modern English. It is substantially more challenging to examine variation and choice in numerical expressions in premodern literate traditions where the survival of texts is highly fragmentary and the entire corpus of materials is orders of magnitude smaller than for English. This brings us to a set of key analytical questions. First, what options exist in each society, or in particular literate traditions, for writing numbers? Second, how does this variation in numerical expression correlate with genre, audience, and medium—in other words, what might help explain the variation? Third, how can we use this variation to usefully explain writers' choices?

Let's start, then, with the recognition of variation—things are not always the same. In nearly every literate context, number is expressible using a variety of strategies, as we've just seen. But just describing variation isn't enough—as social scientists and humanists we should be trying to explain variation. In using the concept of *agency*, I am stepping into tricky territory, because of the often-interminable and unproductive social scientific discussions over the past decades about what constitutes agency, how it relates to social structure, and other theoretical quagmires (Giddens 1979; Ahearn 2001; Dobres and Robb 2000). But put simply, agency is the ability to make meaningful choices among options. In other words, I want to draw our attention not only to what sorts of notational options are available to writers, but also to how writers select among them. Variation can be a key or a clue to the existence of agency, but they're not the same thing.

A lot of the scholarship on numeration and number systems doesn't talk about either variation or agency. Part of the problem is that we—and by "we," I mean both specialists in number systems as well as people more generally—have tended to think of numbers as a technical solution to problems, as a sort of adjunct to mathematics. As I've outlined earlier, I see numerals as representational systems, related to practices of literacy and writing, not as computational systems. They have consequences for numerical cognition, yes, but their use and history are not driven by those consequences. The scholarship on literacy and writing systems is full of discussions of agency (Englehardt 2012). By drawing as much attention to variation and agency in the writing of numbers as has already been done for writing in general, I think we can get to a more satisfying account of how writers chose and used the notations available to them.

When we're talking about choice in numerical systems, we need to use another concept, this one from semiotics, the concept of *modality*, which derives from the philosopher of language Charles Sanders Peirce and has made its way into a lot of theory in semiotics and the social sciences. Modality includes the medium in which something is expressed—visual, auditory, etc.—but also involves the mode of its expression—how it conveys the meaning, and the degree to which it claims to be a representation of some underlying reality. So, for instance, the concept "stop" may be expressed in speech, in writing, gesturally by an outstretched hand, or simply through the ideogram of a red octagon (figure 6.4). I guarantee you will stop your car at a red octagon, even with nothing written on it. Please, please stop at a plain red octagon, just in case. But modalities can also be combined—a red octagon combined with the written word STOP, or a forward-facing palm combined with a shouted "STOP!" Both the octagon and the written word are visual, but the modality is different. In combination, these are *multimodal*

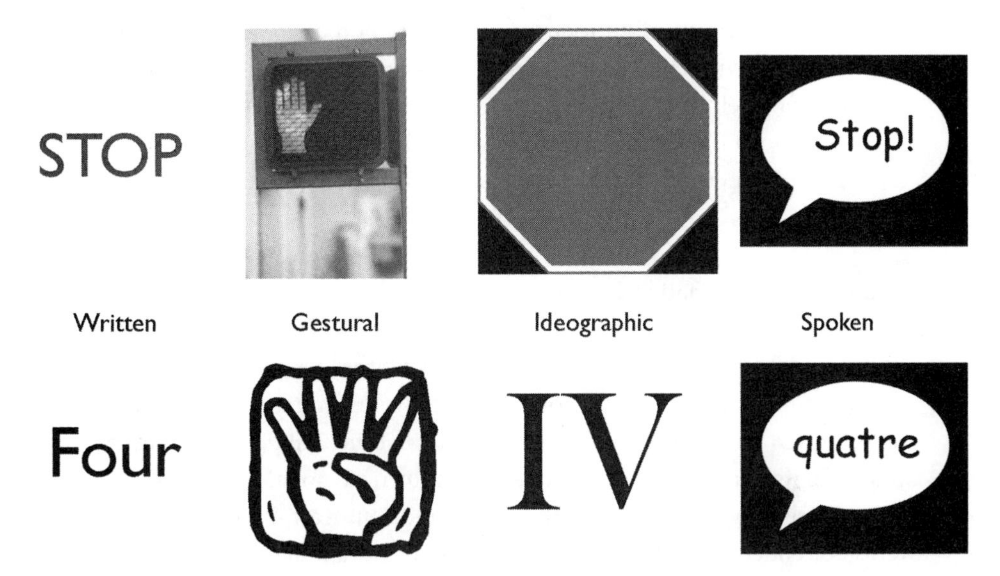

Figure 6.4
Modalities of representing "stop" and "four"

representations, which achieve more together than would be possible as the sum of their parts (Hull and Nelson 2005). As the Sumerologist Cale Johnson notes, "each modality acts as an implicit metalinguistic confirmation of the meaning of the other." Particularly when modalities are brought into coherence with one another, the effect is to encode meaning and simultaneously signal something about the choice being made (Johnson 2013: 28).

All of these same modalities are found in number systems: the written or spoken word "four," a hand with four fingers extended, or a numeral IV. Here too, multimodal representations are available, as in the expression 1.2 million—numerical notation and written number words are two different modalities, within the visual medium, that frequently co-occur and work together to convey meaning. We can then call these multimodal numbers—ones that combine modalities in various ways. When they work together, we may not even notice this consciously. But they provide writers with opportunities to highlight particular readings, make particular numerals more salient, or facilitate more rapid reading—in other words, to use the text to facilitate a particular communicative relationship with some audience.

Below, I set out a model for describing and analyzing variation in multimodal numerical expressions in a comparative perspective. In particular,

I look at two modalities (introduced in chapter 1) that are extraordinarily common cross-culturally: lexical numeration—number words—and numerical notation itself. While there is no one reason why writers choose to mix these two modalities, it happens so often, and in particular ways, that it is no mere coincidence. So let's have a look at how some written traditions do this.

Agency without variation

At one end of the continuum, there are a few writing systems that have no corresponding numerical notation—in other words, numbers always have to be written out as words. Many of the scripts of the Philippines and Indonesia, for instance, historically used no numerical notation. The scripts from which they are descended, the Indian and Southeast Asian writing systems descended from the Brahmi script, had or have numerical notation—ultimately that's where Western numerals came from. But in most of island Southeast Asia, numbers were traditionally written in words, never in numerical notation. It should not be entirely surprising that some script traditions have no numerical notation—after all, numbers can always be expressed in the numeral words of some language, so in some sense numerical notations are redundant. Their worldwide near-ubiquity is thus all the more striking.

In direct contrast, most of the earliest script traditions worldwide use numerical notation but rarely number words. Numbers are widespread in the earliest texts of Mesoamerica, China, Egypt, and Mesopotamia, the four regions of the world where writing is generally thought to have developed relatively independently. In all of these early scripts, numerical notation arrives early, and numbers are not written out in words normally or at all (Chrisomalis 2009). Houston (2004: 238) contends that "most early scripts use word signs bundled with systems of numeration that probably had a different and far-more-ancient origin." We don't know much about Neolithic and other notations prior to writing—most of this evidence simply hasn't survived (Postgate, Wang, and Wilkinson 1995). Whatever the case, it's clear that there is some connection—numerical notation immediately precedes writing and maybe even causes writing to emerge, although the latter claim is much harder to sustain (Schmandt-Besserat 1992).

Egyptian presents an extreme case because of the near-absence of numeral words across its many thousands of texts. A few Old Kingdom hieroglyphic Pyramid Texts (ca. 2350–2100 BCE), written on the sarcophagi and walls of the pyramids of the pharaohs of the period, use both number words and number symbols side by side. One striking example is discussed by Pascal Vernus (2004: 284) where the phrase "nine bows" (*psḏt pḏwt),* a collective term referring to non-Egyptian lands, appears in three different places in different Pyramid Texts in three different forms (figure 6.5)—once with only nine vertical strokes for 9, once with nine bows each drawn separately, and once with both nine strokes and the word "nine" written out phonetically, side by side. These aren't exactly multimodal texts—they all occur in the same genre, the Pyramid Texts, using the same phrase, but were written at different times by different writers. These cross-textual correspondences do, however, allow us to know how the numerals were pronounced (minus the vowels, as with Egyptian writing systems).

Beyond these few texts, though, almost all of what we can reconstruct of the Egyptian number words comes from Coptic texts from the early first millennium CE, which help us fill in vowel sounds and reconstruct the number system more fully (Loprieno 1995). Coptic is clearly an Egyptian descendant language, but its script is largely Greek-derived with only a few signs taken from Egyptian demotic writing (Loprieno and Müller 2012). The hieroglyphic script, used primarily in monumental contexts, the hieratic script used cursively to write on papyrus, and the demotic script used in the first millennium BCE and later all use numerical notations virtually all the time, which presents a real challenge in linguistic reconstruction. Thus,

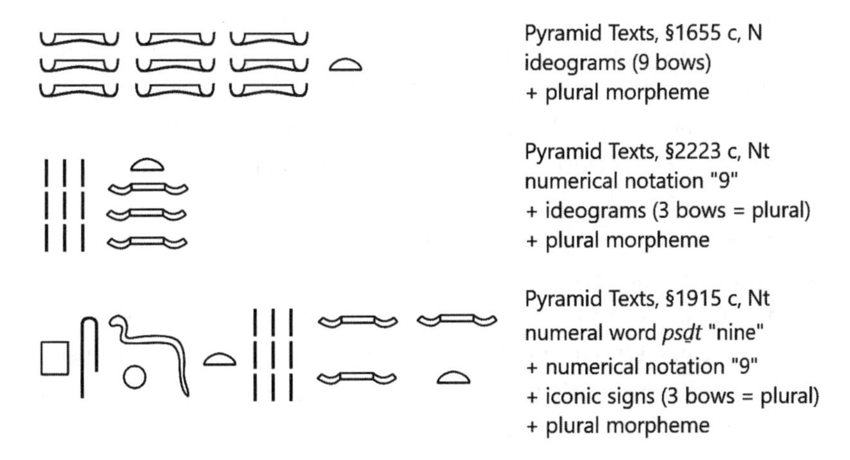

Figure 6.5
Egyptian expressions for "nine bows" in three different Pyramid Texts (after Vernus 2004: 284)

James Allen's (2000: 99) magisterial grammar of Middle Egyptian advises students (correctly) that "it is not necessary to learn all these number words in order to read hieroglyphic texts"—which does raise the question of how Egyptology students ought to read them, when they're reading Egyptian. Mostly, I am told by several graduate students in the field, they just read them out in English or skip over them.

One is struck by this absence in the enormous Egyptian corpus of tens of thousands of texts, across three different writing systems, over three thousand years. It is hard to believe that writing numerals out in words would never have occurred to scribes. The Pyramid Texts show us that they could, on occasion, do so. Rather than sitting there bemused, we should then ask whether some other principle was more relevant. I suspect the norm at work was part of the graphic norms of the very conservative scribal practice, which the Egyptologist John Baines calls "decorum" (Baines 2007). Baines contends that in many respects, Egyptian art, iconography, and written practice weren't so much blindly traditionalistic as they were consciously fixed in place by a set of canons. In other words, it wasn't that Egyptian scribes couldn't have written numbers in words, or that they never thought of it, but that they *chose* not to, consistently, over a period of three thousand years. This demonstrates, moreover, that variation and agency are not the same. Sometimes a lack of variation where we might expect it can also be a sign of agency.

Linear B, the script of Mycenaean Greece and the Aegean islands, is another interesting case where there is a near-absence of lexical numerals in the thousands of clay tablets and other surviving texts. We can infer what the archaic Greek numeral words probably were, working backward from later Greek texts and using comparative historical linguistics, but they were virtually never written out in the Linear B inscriptions. Instead, very simple abstract graphic numerical notations—lines and circles—were common to both Minoan and Mycenaean notational practices. In this case, the exception is all the more interesting for its relevance to the decipherment of Linear B. In his famous book on the decipherment, John Chadwick, one of its co-decipherers, reprinted a letter from May 1953 from Carl Blegen to Michael Ventris (now recognized as a primary figure in the decipherment), writing in excitement about tablet P641 from Pylos:

> Enclosed for your information is a copy of P641, which you may find interesting. It evidently deals with pots, some on three legs, some with four handles, some

> with three, and others without handles. The first word by your system seems to be ti-ri-po-de and it recurs twice as ti-ri-po (singular?). The four-handled pot is preceded by qe-to-ro-we, the three-handled by ti-ri-o-we or ti-ri-jo-we, the handleless pot by a-no-we. All this seems too good to be true. Is coincidence excluded? (Chadwick 1990: 81)

Shortly after Ventris's initial identification of syllabic values for the signs that would secure the language of the tablets as Greek, Blegen had noted the correlation on P641 of words containing the morphemes *ti-ri* and *qe-to-ro* with pictographic representations of vessels with three feet, or three or four handles, as seen in figure 6.6. These words are exactly what we would expect in archaic Greek on the basis of linguistic reconstruction. P641 actually

PYLOS TABLET N° 641·1952

TI-RI-PO-DE

TI-RI-PO

QE-TO-RO-WE

TI-RI-O-WE

QE-TO-RO-WE

TI-RI-JO-WE

Figure 6.6
Linear B tablet from Pylos, P641, with different numerical modalities (drawing by Michael Ventris; image courtesy of the University of London Institute of Classical Studies Ventris Archive (MV 062.3))

contains three different modalities of numerical representation: lexical number words (the numerical prefixes used to indicate the values 3 and 4); pictograms with numerical indicators of 3 or 4 feet or handles; and finally, the ordinary Linear B numeral signs, indicating the quantities of each type of vessel using vertical strokes.

But rather than thinking solely about this tablet's utility for the modern decipherment of Mycenaean Greek, I want to draw our attention to its graphic complexity as an index of scribal choice. For a scribe to represent the number of handles or legs on a vessel in two different ways—both lexically and pictographically—is a significant choice on small clay tablets such as these, where space is at a premium. The writer may have felt it useful, or thought it essential, to employ both forms, perhaps for less-than-fully-literate readers who may not have known words like ti-ri-po-de—even though the modern reader might see the word *tripod* jumping off the page, just as Blegen did in 1953.

Finally, there can be variation in numerical expressions even within numerical notation itself. In Western mathematics, for instance, one can write the same number as a fraction (1/4) or a decimal (0.25), or one can write a number with auxiliary marks like commas and decimal points, or not (1200 vs. 1,200 vs. 1200.00). All of these are numerical notation, but one still has choice as to which one to employ. In some script traditions where there is a close correspondence between signs on the one hand and words/morphemes on the other, it is not even always possible to unambiguously distinguish numerical notation from number words, such as the various Siniform writing systems of East Asia. Whereas in English we have both 4 and *four*, in Chinese there is, normally, only 四. The distinction between written number word and number sign breaks down when each numeral word has a single sign. But this is not to say that there is no variation in writing practice. Figure 6.7 shows several Chinese representations for the number 20,406.

The first, and earliest, is an example of what I have described in chapter 1 as a multiplicative-additive system—it uses signs for 1–9 and for each power of 10, combined in multiplicative pairs, so that 20,406 is written 2 10000 4 100 6. From the Shang Dynasty onward, that was the normal way to write the number. But starting in the late sixteenth century, a sign for zero, *líng*, could be inserted (and read) to indicate empty positions, as in the thousands and tens place in 20406. *Líng* is a different sort of zero,

Classical	二 萬 四 百 六
	2 10000 4 100 6
Classical (with *líng*)	二 萬 零 四 百 零 六
	2 10000 0 4 100 0 6
Positional (with *líng*)	二 零 四 零 六
	2 0 4 0 6
Positional (with zero)	二 〇 四 〇 六
	2 0 4 0 6
Financial (*dàxiě*)	贰 萬 肆 佰 陆
	2 10000 4 100 6

Figure 6.7
Variation in Chinese numerical expressions

though—when used in the classical system, it's really redundant, because the signs for the powers are still included. This also meant that, in many numeral phrases with successive zeroes, only one *líng* was used, so that 20,006 would just be 二萬零六 (2 10000 0 6). Finally, today Chinese numerals are frequently written positionally, using either the *líng* sign or a circular zero, but without the multiplier signs for the powers. All of these are read, character by character, in different ways. Today in China, it is just as common for writers to simply use the Western signs for 0 through 9 outright. There is also, in common use in finance, a secondary set of numerals, the *dàxiě* (lit. "big writing") numerals. These are read identically to the ordinary classical numerals but exist for banking purposes, with signs specifically selected for their complexity in order to prevent fraudulent alteration. This practice is directly analogous with the security measure of writing numerals both lexically and graphically on Western-style checks. Finally, Chinese has a variant sign for 2 (兩), pronounced *liang* instead of the more common *èr* (二). It may seem odd to have two words for *two*, but thinking about the English lexicon for a moment, and considering the "twoness" of *couple, second, pair, deuce, duo, twain, twice,* and *both,* it is not so surprising after all.

Thus, even in contexts where there is seemingly only one choice—where numbers are only written using one modality—there is often variation available to the writer. Now let's turn from cases where numbers are expressed typically in only one modality to ones that mix modalities in various ways. Three such ways are:

1. Blended modalities: numerical notations that provide cues or clues to their pronunciation or their linguistic origins;
2. Hybrid modalities: numerals expressed through a combination of lexical and notational resources, but with the two remaining distinct;
3. Parallel modalities: numerals expressed in two or more modalities within a single text.

Blended modalities

Sometimes the distinction between lexical and graphic modalities is not so clear-cut. Normally, numerical notation is translinguistic and can be read in different ways in different languages—in other words, there is no phonetic component to it. Sometimes, however, a numeral phrase is written using signs that provide some clue to the phonetic value of the relevant number word, but without using full representation in words. I will call these *blended modalities*.

Immediately your mind might turn to the Roman numerals, where C is the first letter of *centum* (100) and M is the first letter of *mille* (1,000). However, this was actually a later development. Note that the other Roman numerals don't have any phonetic associations, even though they are letters—why would L be 50 or X be 10? As mentioned in chapter 3 (and simplifying a very complex story), the Roman numerals began as a completely different sign system from the Roman alphabet, one that was not associated with letters at all; over time, the signs became associated with letters which integrated them with literate practices but also obscured their origin (Keyser 1988). By coincidence, the old Roman Ɔ for 100 could be mirrored across the vertical axis to become C, and the Roman ↀ for 1,000 could be separated at the bottom to resemble an M. Actually, ↀ was the main form of 1,000 in antiquity, and even throughout the medieval and early modern period it was extremely common.

The same sort of assimilative process went on with the closely related Greek acrophonic system. Originally unconnected to the sounds of words, by the sixth century BCE the signs were altered to correspond with the first sounds (Greek *akros* = "first, highest" + *phone* "sound") of the words PENTE, DEKA, HEKATON, CHILIOS, and MYRIOS as their respective numeral signs for 5, 10, 100, 1,000, and 10,000 (Tod 1979). Thus, for instance, 135 would be HΔΔΔΠ (100 + 10 + 10 + 10 + 5). But note that when written out as such, Greek acrophonic numerals cannot be "read" as words: the ancient Greek word for 30 is *triákonta*, not *deka deka deka*. Moreover, the sign for 1 (a vertical stroke) is not acrophonic at all. At best, the acrophonic numerals probably helped in learning the signs and, for semiliterate readers, drew associations that made reading easier. Blending modalities served a mnemonic function, rather than one oriented toward the needs of fully literate readers.

Acrophonic numeration survives today in Western notations such as the use of K for 1,000, as in popular video games like *NBA 2K18*, or in *Y2K*. The K is the first letter of the pseudo-Greek morpheme *kilo*, which was stripped in the early modern period of the proud chi at the start of χίλιοι (or, I must add indignantly, *Chrisomalis*) when it became a metric prefix (as in *kilogram*). Now, over the past few decades, K has turned back into an acrophonic numeral for 1,000. According to the *Oxford English Dictionary*, K for 1,000 seems to have originated in the 1960s in computing and electronic contexts (where it normally refers to 1,024 bytes, not 1,000), then spread to salary figures like "She makes $92K a year" in the 1970s. Since Y2K, K has become a playful and hypermodern (even though millennia-old) strategy for representing years. It's no shorter to write 2K18 than 2018, and presents uncertainty to readers as to how it should be read aloud, even where its meaning is unambiguous. Rather, it's an example of what the linguist David Crystal (2001) calls "ludic language"—language influenced by the human capacity for play, manipulation, and puzzling.[2]

Another case of blending modalities is the *siyaq* or *dewani* notation used by Arabic, Persian, and Ottoman administrators from the tenth to the nineteenth centuries (figure 6.8). Here, the sign for each multiple of each

LES CHIFFRES SIYÂK DANS LA COMPTABILITÉ PERSANE

CHIFFRES	VALEUR en mann (1)	CHIFFRES	VALEUR en kharvârs (1)	CHIFFRES	VALEUR en kharvârs (1)
[illegible]	1	[illegible]	1	[illegible]	100
[illegible]	2	[illegible]	2	[illegible]	200
[illegible]	3	[illegible]	3	[illegible]	300
[illegible]	4	[illegible]	4	[illegible]	400
[illegible]	5	[illegible]	5	[illegible]	500
[illegible]	6	[illegible]	6	[illegible]	600
[illegible]	7	[illegible]	7	[illegible]	700
[illegible]	8	[illegible]	8	[illegible]	800
[illegible]	9	[illegible]	9	[illegible]	900
[illegible]	10	[illegible]	10	[illegible]	1,000
[illegible]	20	[illegible]	20	[illegible]	2,000
[illegible]	30	[illegible]	30	[illegible]	3,000
[illegible]	40	[illegible]	40	[illegible]	4,000
[illegible]	50	[illegible]	50	[illegible]	5,000
[illegible]	60	[illegible]	60	[illegible]	6,000
[illegible]	70	[illegible]	70	[illegible]	7,000
[illegible]	80	[illegible]	80	[illegible]	8,000
[illegible]	90	[illegible]	90	[illegible]	9,000

Figure 6.8
Persian variant of *siyaq / dewani* numerals (Kazem-Zadeh 1915: plate VIII)

power of the base of 10 is a cursive reduction of the corresponding Arabic number word. This is like a visual clue indexing a particular linguistic reading, as if the number 9 looked like the word *nine*. Yet despite their lexical origin, *siyaq* numeral phrases could not simply be read lexically. They are too reduced to be read or written automatically except by a trained scribe, and constructing numeral phrases using two or more *siyaq* signs does not automatically generate a grammatical Arabic number word. There is also no necessary expectation that a scribe using *siyaq* numerals would have been knowledgeable in Arabic, and we know that a lot of them were Turkish, Farsi, or Hindi speakers. Figure 6.8 shows the Persian version of the *siyaq*, and while some of its users may have known some Arabic, that would not have been necessary. The origin or the "visual etymology" of the signs is lexical, but their reading and use are not necessarily so. Actually, all users of the *siyaq* numerals were also fluent users of other numerical notations, such as the Arabic, Persian, Indian, or Western variants of the decimal, positional numerals 0 through 9. The role of the *siyaq* in its social context was actually a form of social control, limiting access to financial information to those initiated in its use, while providing a visual clue that might have been useful in teaching. Blending modalities, in this case, was semicryptographic and part of a set of accounting practices designed to limit the flow of information rather than to encourage it.

A final instance of blended modalities consists of numerals used in unexpected contexts for nonnumerical, often playful purposes, including phonographic ones. In these, rather than the numeral giving a clue to the numerical value, the numeral indexes a phonetic value. Probably the best-known numerical phonogram in English is K-9 used as a pun on the word *canine*, but it is not the only one—2 for *to*/*too*, 4 for *for*, and 8 for *-ate* in phrases like "Stop h8." The use of the numerals for their phonetic value remains frequent, although less so than a decade ago, in the written language of youth texting, much to their elders' dismay.[3] But this is no mere modern trick but a longstanding practice in multiple script traditions. Take, for instance, the Maya numeral 4, expressed as four dots, with the phonetic value *kan* or *chan*. The same four dots can be employed nonnumerically to express homophones or near-homophones in words meaning "snake" or "sky"—the numerical sign could be used for any of the three meanings (Houston 1984). Or, in Elamite cuneiform, the logogram for "king" is the numeral phrase "180 20," or 3,600 (*šar*), which is nearly a homophone of the word for king, *šarru* (Nougayrol 1972; Biggs and Stolper 1983). B4 we dismiss "kids these days" for what we suppose are their linguistic atrocities, we should look to past scribal practice for evidence of numerical play.

Hybrid modalities

Other numerical traditions combine lexical and nonlexical graphic signs in a systematic way, normally by using lexical terms for powers of the base in combination with graphic signs for low numbers. I call these *hybrid modalities*—there are two modalities, but they're not blended together so much as interwoven. Both the lexical and nonlexical parts of a hybrid phonographic representation are discrete and comprehensible on their own, and then they are combined in some systematic way that is directly readable.

A well-developed numerical tradition that uses hybrid phonography systematically is the Eblaite script found on thousands of tablets at the palace archive of Ebla in what is now northwestern Syria, dating to the middle of the third millennium BCE. The Eblaites were speakers of a Semitic language related to Akkadian and Babylonian, and used a cuneiform script akin to those used in Mesopotamia, to the east. For numbers below 100, Eblaite used the ordinary Sumerian curviform or cuneiform signs for 1, 10, and 60, written additively just like almost any other Mesopotamian numerical system at the time (Pettinato 1981; Chrisomalis 2010: 245–247). Where Eblaite diverged from the practice of Akkadian scribes is in the writing of numbers for powers of 10 starting with 100. For 100, 1,000, 10,000, and 100,000, Eblaite did not use graphic signs but represented the powers of the base lexically, using the numeral words. In some cases, these could be reduced to the first syllable of each word (*mi, li, ri, ma*).

In figure 6.9, we have an Eblaite clay tablet probably from the twenty-fourth or perhaps the twenty-third century BCE (ARET 02, 20; CDLI P241045; cf. Edzard 1981). We'll focus only on the bottom left register. The coefficients of each power are easily readable by the number of signs: 1, 8, 6, 2. The words that follow each number are the Eblaite words for each power of ten descending from 100,000 to 100, giving the final total of 182,600. This has a close parallel with the expression "1.2 million" that I discussed earlier. Eblaite didn't have any known numerical signs for the higher powers, even though many of the other Mesopotamian cuneiform writing systems did.

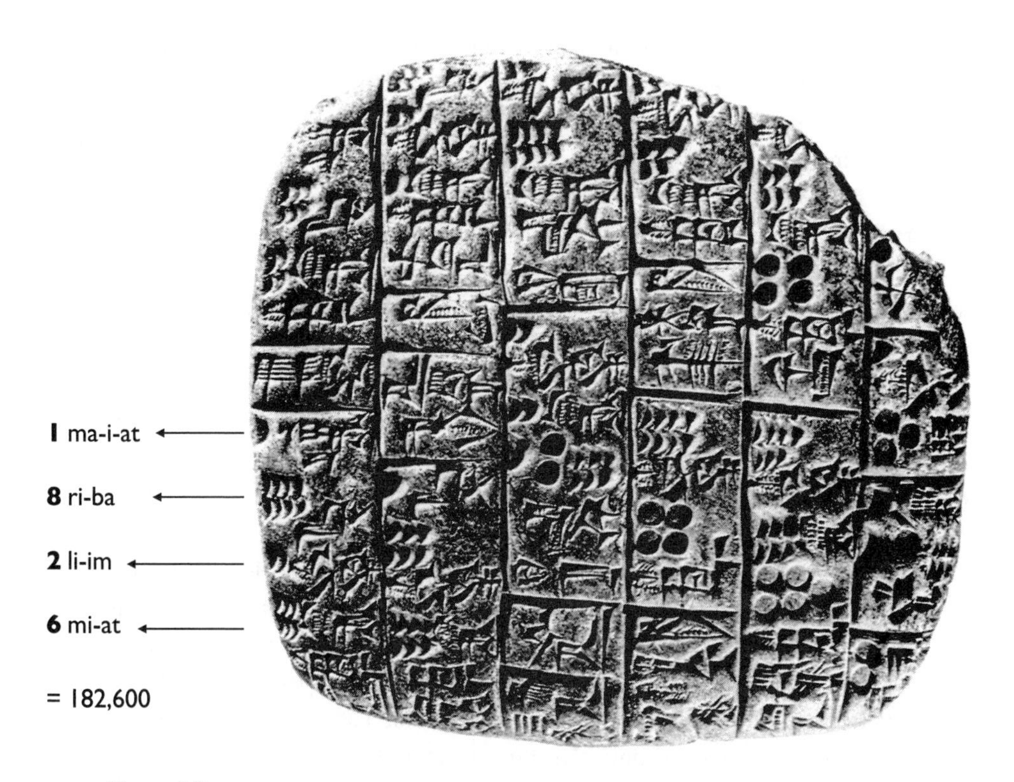

Figure 6.9
Eblaite text with a representation of 182,600 using numeral signs with words (source: Edzard 1981, ARET 02,20)

However, the scribes were familiar with other ways of writing, such as the Sumerian numerals used by both the Sumerians and later, the Akkadians, throughout the third millennium BCE.

Why, then, did they diverge from the Sumerian, sexagesimal numerical practice? In these cases it's often tempting for social scientists to invoke concepts like identity to explain variation, but in most other respects, Eblaite had many similarities with Akkadian scribal practice; simply invoking identity seems like a weak explanation, given that there's no real evidence for or against it. An alternative account might focus on cognitive effort. Sumerian written numerals have a complexity in that they use many elements of sexagesimal (base-60) notation, whereas Eblaite, like all the Semitic languages, is decimal. To write 182,600 in Sumerian, the scribe would need to write five signs for 36,000, four signs for 600, three for 60, and two for 10. It's not that it's longer to write—14 signs is not so many, compared to the text we're looking at—but it requires a great deal of calculation to figure out how to write it in this system, for speakers of a language with decimal numerals, like Eblaite. Analogously: imagine if every time you needed to write out a time in hours, you needed to figure out how many seconds it is first. In contrast, the Eblaite text, although not exactly concise to write unless the words are reduced to their first syllables, requires no particular cognitive load—simply append the numerals for 1 through 9 to whatever word for a power of 10 you need. But the fact that sometimes the short forms were used solves even that problem—note the parallel with the Greek acrophonic system, using the first letter to represent the whole word. Because Ebla was politically distinct from the Akkadian Empire to the east, their scribes may have been more free to experiment. Eventually, the Semitic-speaking Babylonians around 2100 BCE started using decimal numerical notation much like the Eblaite system, alongside a new notation—positional, base-60 numerals that were the earliest true place value system (Proust 2009; Ouyang 2016). The decimal system began to use *me* and *li-im* for 100 and 1,000, perhaps in emulation of Eblaite practice but just as likely independently developed. And, by the seventeenth century, the Hittites had begun using the same Semitic words for 100 and 1,000, along with elaborated ideograms for 10,000 and (probably) 100,000 (Hoffner 2007). Even though Hittite was an Indo-European language, its numerals (like those of most Indo-European languages) were decimal, so this was an easy adoption of Mesopotamian practice, though the signs were probably read aloud in Hittite, not Babylonian or another Semitic language. Sumerian—known and available, though deeply archaic at the time—was not considered.

The Sogdian written tradition of central Asia was used roughly from the fourth to tenth centuries CE in parts of what is now Uzbekistan, Tajikistan, and western China. Sogdian is a highly cursive alphabet, difficult to read today as it surely must have been at the time. In figure 6.10, we have a Sogdian financial document from the eighth or ninth century (Bi and Sims-Williams 2010). The original findspot of the text is unknown, but it is from western China, near the city of Hetian (ancient Khotan). In Sogdian texts, there were both numerical notations and numeral words for all the numbers at least as high as 10,000, so in theory, any text could be written entirely in words or entirely in notation. However, many texts, like the one shown here, use words for the lower numerals one through nine, combined with notation for the powers for 100, 1,000, and 10,000.

Figure 6.10
Sogdian financial text, eighth–ninth century CE, GXW 04320 (Bi and Sims-Williams 2010: 501; Museum of Renmin University of China, Sogdian document no. 3; Image © Bi Bo and Nicholas Sims-Williams)

δs 1,000 pny βyrt **ctβ'r** 1,000 '**βt** 100
ten 1,000 pny received four 1,000 seven 100
"10,000 pny. Received: 4,700 [pny]."

δ'rt 'ytxw msyδr '**δw** 1,000 '**δwy** 100 pny
has Itkhu priest two 1,000 two 100 pny
"Itkhu the priest has 2,200 pny."

δ'rt βwγδ't 24 1,000 pny
has Vogh-dhat 24 1,000 pny
"Vogh-dhat has 24,000 pny."

δ'rt ypγw ''tryc **pnc** 1000 '**βt** 100 pny
has yabghu Atarich five 1,000 seven 100 pny
"The *yabghu* Atarich has 5,700 pny."

This is the exact opposite of the pattern we saw in Eblaite—here the powers tend to be in numerical notation, and the units in words. For instance, in line 2, the number 2,200 is written with two in words, but 100 and 1,000 notationally. But immediately after this, in line 3, 24,000 is written entirely in notation: the sign for twenty, plus four strokes for one, followed by 1,000. A Sogdian sign exists for 10,000 that the scribe could have used if they had wanted to write "two 10,000 four 1,000," but it is rare, and the scribe may not have known it or wanted to avoid it. In any case, in line 4 the scribe returns to the hybrid modality "five 1,000 seven 100" for 5,700.

Again, we can ask why. Here I think it's possible that the decision was motivated by issues related to multilingualism. The numbers one through ten are among the first words learned by children and new learners. This is an economic document, located in or around Khotan, a major strategic location along the Silk Road, and Sogdian was a minority language here, well to the east of its heartland. It's not that the numerals 1–10 would have been hard to write—they're mostly vertical strokes, ligatured together at the bottom, as in the 4 in 24 in line 3. It's that the words for 100, 1,000, and 10,000 couldn't be presumed to be known by all potential readers of the text.

Hybrid modalities are common cross-culturally. Returning to the Roman numerals, in a lot of early modern writing, Roman numerals and numeral words could be mixed fluidly. One problem faced by medieval writers was that there was no standard way for writing large numbers in Roman numerals. Charles Burnett (2002) discusses the interesting case of the *Helcep Sarracenium* of Ocreatus, a twelfth-century text explaining place value and Indian/Arabic arithmetic (the "algorism") including transliterations from Roman to very early Western numerals. This is a multicultural document influenced by Indo-Arabic knowledge but using the representational systems available to a medieval European writer, well before Leonardo of Pisa's *Liber abaci* of 1202 made the Western numerals familiar (to scholars, at least). Because there were no standard ways of expressing numbers higher than 1,000,000 in Roman numerals (the bar or *vinculum* representing multiplication by 1,000 only took you so far), some other strategy was needed. The solution in this text was to simply add the words *decies* and *centies* after

Table 6.1
Hybrid Roman numerals in the *Helcep Sarracenium*

i	1
x	10
c	100
M	1,000
xM	10,000
cM	100,000
MM	1,000,000
deciesMM	10,000,000
centiesMM	100,000,000

the MM for 1,000,000, as shown in table 6.1. Because the Western numerals are infinite, we don't have quite the same problem, because the pure notational forms 10,000,000 and 100,000,000 exist, but we still usually find *10 million* and *100 million* easier to read and understand.

Similarly, Ford (2018) analyzes numerical expressions in the Middle English verse romance *Capystranus*, dating to the late fifteenth or early sixteenth century and notable for its frequent use of large numbers. This was a poem clearly intended to be read aloud, as was the common practice at the time. The meter of the poem clearly indicates how the numerals must have been read. Of the expressions in the poem, fifteen were in lexical numerals alone, six in Roman numerals alone, and nine used hybrid modalities such as *Syxe and twenty.M.* for 26,000 and *C. thousande* for 100,000. It wasn't that there was one way to write any specific number, though—for instance, 20,000 was written as *xx. thousande, Twenty.M* (twice), and *Twenty thousande* (twice). Rather, one of the best predictors for the choice of modality was the position of the numeral in the line. Roman numerals never started a line and never ended a line (so as to clearly indicate the rhyming scheme). Seven of the nine hybrid numerals were at the start of the line, with the lexical numeral preceding the Roman numeral, likely to conform to the aesthetic canon of starting lines with words (comparable with the modern norm of not starting a sentence with numeral notation, discussed earlier).

Now, the whole thing could have been written out in words, and this is the choice made by many poets even today. But hybrid modality is far more common than most contemporary scholars acknowledge. Important English printers such as William Tyndale and William Caxton often avoided the use of Roman numeral C and M in early printed Bibles in favor of lexical hybrids such as "ix hundred and xxx yere" (Williams 1997: 11). We even see cross-modality influences such as the use of XX (*vingt*) for 20 in hybrid French texts, as in a letter from King Richard II of England dating from 1392 but referring to events three years earlier, "l'An Mille, CCC, IV XX. & Neuf" (Rymer 1740: 76). Clearly this can only be read in French as *mille trois cents quatre-vingt et neuf* (1389). Not only is there a hybrid modality, but the use of XX as a structuring feature in combination with IV shows a further influence, in this case, from French, where *quatre-vingt* is the number 80 (Preston 1994). Crossley (2013) shows that mixtures of Roman numerals and number words (French, Latin, or other), mixtures of Roman and Western numerals, or annotations of numerals with signs indicating how they were to be read (e.g., a superscript *o* or *mo* in phrases like M° for *millesimo* = 1,000th) were extraordinarily common in late medieval texts. The frequency of this sort of notation, combined with our awareness of just how frequently contemporary writers use phrases like "27 million," has enormous implications not only for how we read texts, but how we explain longer-term processes of change and replacement of notations like the Roman numerals. Because hybrids are common, we shouldn't treat numerical notations as pure pristine objects to be compared to one another, without considering these sorts of fruitful blends.

Parallel modalities

The third type of multimodal numbers I'm interested in are what we can call parallel modalities. These are examples like John Jacob Astor's check from 1792, in which each individual number is written with only one modality, but across a single text multiple modalities are used in parallel for writing the same number. This sort of apparent redundancy is actually very useful for several reasons, as some examples from the ancient world will show us.

The South Arabian script tradition started in the early first millennium BCE, as a variant of the Bronze Age alphabets of the Sinai and the Levant, but quickly took on a distinct aesthetic quality especially in monumental contexts. Used mainly in the southern part of the Arabian peninsula (Yemen and Oman), but also in other parts of Arabia as well as across the Red Sea in what is now Eritrea, it was written in *boustrophedon* style—in other words, with lines alternating right to left and left to right. For the first several centuries of its existence, South Arabian was a writing system without any special numeral signs—numbers were only written out in words. Then, around the sixth or fifth century, possibly under the influence of the Greek acrophonic system I discussed earlier, a set of distinct additive South Arabian acrophonic numerals developed (Biella 1982).

In figure 6.11, we see an inscription from Sirwah, probably from the sixth or fifth century BCE, in which the number 6,000 is written. South Arabian numerical notation is always preceded and followed by a hatched vertical bar—so here, we see the six signs for 1,000 surrounded by bars. This creates a visually salient indicator that makes numerals distinct within the text. What's interesting with South Arabian is that writers almost never used numerical notation alone, but rather, preceded by the exact same number word.[4]

Figure 6.11
South Arabian inscription for "6,000" with parallel modalities (DAI Sirwah 2005–2050)

sdtt 'lfm 1,000 1,000 1,000 1,000 1,000 1,000
"six thousand (6,000)"

One possible reason for this redundancy is that numerical notation was rare in the South Arabian writing system. It hadn't existed for centuries previous to this inscription, and at the time it was fairly new, so writers may not have assumed that even fluent readers would be familiar with it. It's also possible that this duplication serves the same security function as on modern checks, or as in legal contracts where redundant numbers in parentheses are used to ensure correct readings, and I've transcribed it here as if it were so. However, in these sorts of monumental inscriptions, it doesn't seem likely that scribes carving in stone would have this concern. A final, intriguing possibility is that adding numerical notation at all was simply done to emphasize and make salient the number being expressed—in other words, the visually striking repeated signs, surrounded by hatched bars, are part of an aesthetic canon designed as a sort of emphatic flourish. The parallel here is with the "conspicuous computation" strategy I discussed in chapter 2. This sort of monumental inscription is exactly the sort of place where salience—giving prominence to numerals within a text—might be a viable strategy.

In any event, this practice did not survive too long. Although the South Arabian script tradition survived right up until the early Islamic period, the South Arabian numerical notation was not used past the first century BCE. After that point, the script reverted to writing all numbers out in words alone. This process of simply abandoning numerical notation may seem strange, given how essential we, as modern people, imagine it to be to the survival of civilization. But the history of numerical notations is not a unilinear, progressive arc toward representational and computational perfection. Not only will systems be replaced for reasons other than "efficiency"—for whatever value of "efficiency" you might choose—sometimes they won't be replaced at all, but simply abandoned. To me, the abandonment of numerical notation in South Arabian writing just reinforces that it was always an auxiliary system, never central to scribal practice.

A related example of parallel modalities comes from a very interesting Greek text, Fort. 1771, found among the Persepolis Fortification Archive, analyzed and curated at the Oriental Institute of the University of Chicago (figure 6.12). The Archive is one of our most precious bodies of knowledge about economic and social life during the Achaemenid empire centered in Persia but spreading from the Balkans to the Indus Valley. At Persepolis, the ceremonial capital of the empire founded by Cyrus the Great and Darius I in the late sixth century, tens of thousands of economic and administrative texts from around 500 BCE have been found, mostly in Elamite cuneiform and in Aramaic, along with a few unusual texts in other languages. Fort. 1771 is the only Greek text in the Archive, at a time when trade and conflict

Figure 6.12
Fort. 1771, Greek tablet from the Persepolis Fortification Archive, ca. 500 BCE (courtesy of the Oriental Institute of the University of Chicago)

ΟΙΝΟΣ ΔΥΟ II ΜΑΡΙΣ
oinos duo II *maris*
wine two II maris
"two (2) liquid measures of wine"

between Greeks and Persians was common, but not particularly involving Persepolis, which was a new city at the time.

But, as Matt Stolper and Jan Tavernier (2007) rightly note, this is not some misplaced text that just accidentally happened to fall out of some Greek merchant's satchel. It has an Elamite seal on one end, deals with a commodity of wine, and uses the Persian unit *maris* to measure it, in common with many of the other texts of the archive. The language of the text may be different, but the function and the information system are the same. I give here just the text on the obverse of the tablet, "oinos duos II maris." Once you know that *maris* is a unit of liquid measure (equal to about 9.3 liters), you need almost no Greek to read it as two units of wine. As with the South Arabian text from Sirwah, it looks like the number word two is followed by two vertical strokes as a sort of redundancy.

How was this text meant to be read and understood—and how should we think about the choice to inscribe two vertical lines below the Greek word *duos*? Unfortunately, we don't have a Persepolis Manual of Style to tell us how to read and write numbers, and neither did the scribe who wrote it at the time. Here, aesthetic preferences don't seem a plausible explanation—this is not a beautiful display text. In an excellent and underappreciated recent article, Flavia Pompeo (2015) begins with the observation, first made by Schmitt (1989), that the two vertical marks seem to have been added between *duos* and *maris* shortly after the rest of the text had been written, judging by the spacing, to clarify for readers who may not have been fluent in Greek. This seems right—it's on its own line, rather tightly placed between the lines above and below it. But Pompeo then notes that even the two marks are not self-explanatory. Perhaps they are Greek acrophonic numerals: two vertical strokes would be 2 in this system, which I discussed earlier. Or perhaps they are meant to be Aramaic numerals: Aramaic would also use two vertical strokes. Or perhaps they were intended as linear models of the Elamite or Babylonian cuneiform numerals (two lines standing for two wedges). Or perhaps they are just two general tally marks, not really intended as part of a specific notation but intended to be readable by just about anyone. We'll probably never know, but it bears on the question of who exactly added them, and why. Were they added by a thoughtful Greek-speaking scribe as a sort of afterthought? Alternately, perhaps a bilingual Elamite or Persian scribe thought to clarify the text for readers who might not have been familiar with the word *duos*. Perhaps the marks were part of an administrative practice by which the secondary notation was added as a security measure or a kind of reckoning tool for when the two *maris* reached some destination.

One final twist: imagine, hypothetically, that the marks were there but the word *duos* was not. Would we then be so confident that two random lines meant the number 2, as opposed to two instances of the letter I, or some incomplete letter? Epigraphers often struggle to clearly identify marks like these as numerals, because they're open to so many other interpretations. It's probably right to think that the two marks were put there to clarify the *duos*, but their clarity derives in part from the fact that the lexical and notational numerals are both there in parallel, not just one or the other.

One final example comes from the earliest period of the Aramaic script tradition as used in Assyria in northern Mesopotamia (figure 6.13). The bronze Assyrian lion weight BM 91220 from Nimrud found by Sir Austen H. Layard in the 1840s bears an early Aramaic inscription; it probably dates from the reign of the emperor Shalmaneser V (726–722 BCE). The social context of the period was complex and multicultural. At the time, the Aramaic language and script were relative latecomers to the environment, having only recently been elevated to the status of lingua franca of the Assyrian empire under Shalmaneser's predecessor, Tiglath-Pileser III (745–727 BCE). It was widely spoken, but still only tentatively a prestige language. The lion weight, one of sixteen Layard found, is notable in that it bears three inscriptions, two of which are Aramaic and one purely numerical (Fales 1995).

On the right flank, the weight of the object, "15 minas" is expressed in the Aramaic decimal numerical notation: a sign for 10 followed by five signs for 1. Below it, on the right base, the amount is written out in Aramaic words: *khamshat ashar*, five plus ten. Finally, on the left flank, there are fifteen ungrouped vertical strokes, a sort of tally.

What motivated the triplication of the value of 15? Was it in part so that the value could be seen from both left and right sides? Perhaps, but then why write it three different ways? Could it be to prevent fraud, as in check-writing practices? Perhaps, but this is an inscription on bronze, and hardly easy to alter—and in any case, its weight would give the alteration away. More likely, some of the users of the weight would not have been literate in Aramaic, and might not have been able to read either of the inscriptions on the right side. The tallies serve a clear and useful purpose in this case—they are probably readable by just about anyone. Even those literate in Aramaic need not have known 15 as a combination of a horizontal 10 plus

Figure 6.13

Aramaic lion weight, Nimrud, ca. 725 BCE (BM 91220; image © The Trustees of the British Museum)

Right flank:
mnn —||||| b zy `rq`
mina-PL 15 by the land
"15 minas, by the standard of the land"

Right base:
[ḥ]mšt `šr mnyn [b zy] mlk
five ten mina-PL [by the] king
"fifteen minas, by the standard of the king"

Left flank:
|||||||||||||||

five vertical strokes. Aramaic and Phoenician writing had been around for a while, but the lion weight inscription is actually the very earliest example of an unambiguously Aramaic numerical notation. In other words, the numerical notation was a novelty in the newly Aramaeized Nimrud of the mid to late eighth century. Thus, in choosing three modalities—one using the novel Aramaic numerals, a second using Aramaic words readable by anyone literate in the script, and a third using a tallying system accessible to anyone—the scribe was considering the ability of the inscription to be readable by all its potential audiences.

Code choice

So far I've discussed blended modalities, where numerical notation contains a cue or index to a lexical reading; hybrid modalities, where a single number is written using two (or more) distinct notations, one lexical, the other graphic; and parallel modalities, where the same number is written two or more times in different modalities. But what about the simplest case of all, texts that use number words in one set of contexts and numerical notation in another? These we might simply classify, in a sociolinguistic model, as an example of code choice—we select an expression based on some goal, interest, or principle.

Let's return to the sample sentence advocated in the *Chicago Manual of Style* (figure 6.2, above), the almost-baffling "Five hundred and ninety-three men, 417 women, and 126 children under eighteen, besides 63 of the crew, went down with the ship." In this example, "Five hundred and ninety-three" starts the sentence, and the norm is that sentences start with words, not numerical notation. One of the rationales for this principle, whether the original explanation or not, is that sentences should start with a capital letter, as a way of visually marking sentences, and, after all, there are no capital numbers. In contrast,[5] 417 and 126 are three or more digits so should be written with numerals, and while 63 should be written out in words by the style guide's rules, keeping it in the same style as the previous two figures of persons makes some sense. Eighteen, however, should stay as a word because it's a relatively low number, or perhaps because it's an age rather than a count of people. You may disagree with any of these decisions and might have made them differently. That is the whole point of an agency-based account of numerical variation. But the writer undeniably expected all fluent readers of English to be able to quickly read and understand the sentence. However, if you asked the average reader thirty (or 30) seconds later, they likely couldn't tell you which was expressed in which format.

The Etruscan numerals provide a fascinating case of variation in code choice in part because even though Etruscan script is readable, the Etruscan language is poorly understood due to a lack of known relatives; but numerical variation in texts helps us to improve that understanding. It's not an Indo-European language like Latin and the other Italic languages, but coexisted on the Italian peninsula for centuries until finally going extinct probably in the first century CE. The Etruscan numeral words, in particular, have been the source of some perplexity for decades because in the absence of clear context, it is difficult to assign specific numerical meanings to specific words. The problem is made even more extreme in that, of around 10,000 extant Etruscan inscriptions, around 200 write numerals using a numerical notation very similar to, and in fact ancestral to, Roman numerals, which is completely understood, but only around 40 known inscriptions appear to have numerical values (often ages of deceased individuals) written using number words (Kharsekin 1967). Because the numerical notation can be read precisely, and because many of the numerals designate age at death on tombs and sarcophagi, Kharsekin (1967) a half-century ago was able to evaluate most of the lexical numerals by correlating "age curves" of the graphically inscribed tomb texts, whose reading was secure, with those with lexical numerals, which were potentially ambiguous. But we don't know why writers chose lexical expressions in around 15–20% of the available inscriptions. The variation here is useful to us, as modern decipherers, but otherwise not really analyzable.

The chief remaining challenge was that, while the numerals for *four* and *six* could be identified as *huθ* and *śa*, which was which remained in dispute, with important Etruscologists on either side of the issue. In 2011 a study in the journal *Archaeometry* shed light on the issue using variation across modalities in a very unusual corpus of "texts": Etruscan six-sided dice (figure 6.14) (Artioli, Nociti, and Angelini 2011).

Of 93 Etruscan surviving dice, 91 have one through six pips on the faces, as is typical of modern dice, but two very remarkable examples, instead, use the first six Etruscan numeral words. For the 91 dice with pips, there were only two patterns by which the numbers were distributed on opposite

1	θu	⚀	I
2	zal	⚁	II
3	ci	⚂	III
4	śa	⚃	IIII
5	maχ	⚄	Λ
6	huθ	⚅	IΛ

Figure 6.14
Etruscan numeral words, dice, and numerical notation

faces. Dice where 1 opposes 2, 3 opposes 4, and 5 opposes 6 were more prominent in dice from the eighth through fifth centuries BCE, while others (as in modern dice) opposed 1 to 6, 2 to 5, 3 to 4, typically later examples from the fifth through third centuries. But note that in both of these formats, 3 opposes 4, and in both of the two dice with lexical numerals, *śa* opposes *ci* (three), so *śa* is overwhelmingly more likely to be 4 rather than 6, and *hu*θ must be 6 rather than 4. Representational variation allowed the solution of a century-old linguistic problem.

Earlier, I discussed the Greek acrophonic numerals. However, the acrophonic numerals were used only for a small range of domains and functions in antiquity (Tod 1979; Threatte 1980). For instance, they were never used to express ordinal numbers, only cardinal values. They were never used to express dates, or the ages of the deceased in funerary inscriptions. They were virtually never used in connected prose at all, such as decrees. In these and other contexts, numerals were expressed lexically. There was nothing to prevent acrophonic numerals from being used in these contexts, except, perhaps, the fact that numbers, like texts themselves, were generally read aloud at this period; i.e., silent reading was not the norm in classical antiquity (Saenger 1997).

After about 325 BCE, another numerical notation became common: the alphabetic numerals, in which the 24 letters of the alphabet plus three additional letters were assigned the values of 1–9, 10–90, and 100–900 (Tod 1979; Johnston 1979). Alphabetic numerals had been around for several hundred years, having an ultimate origin in seventh-century BCE contact with Egypt (Chrisomalis 2003). But this did not affect the widespread use of lexical numerals in connected prose. What it did do, though, was encourage the Greek writing of number words in descending order in the units and tens (compare "twenty-four" with "four-and-twenty"). Prior to around 325 BCE, descending order from the highest to the lowest power was rare; afterward, it was common, a fact that Keyser (2015) correctly attributes to increasing numeracy and familiarity with numerical notation in that period. In other words, as writers began using numerical notation more frequently alongside lexical numerals, linguistic structures that were consonant with those systems increased, and ones that contradicted them decreased. There is nothing requiring such a transition—German, to this day, still places units before tens in its number words, but no one seriously thinks that German speakers are less numerate than others. But it illustrates the ways in which numerical modalities, even when not used for the same purposes or in the same contexts, can influence one another.

In chapters 3 and 4, I discussed the transition from Roman to Western numerals in terms of an extremely gradual, socially motivated replacement influenced by the increase in literacy associated with printing practices. We saw that Western numerals were adopted in some rather haphazard ways throughout early printed books, and not in ways that necessarily fit the model of rationality often presumed to be the case. This situation clearly fits the model of code choice. To this, we then need to add the abundance of texts that use Roman numerals along with alternatives such as Western numerals or number words for different purposes. This will help us to think of the rise of Western numerals not only in the context of the Roman numerals—their purported competitor—but also in the context of the lexical numerals, which never went away.

Nor did the mixing of numeral systems end with the so-called transition to Western numerals. A South Carolina eight-dollar bill from 1776 (figure

6.15), from the same generation of early American life when John Jacob Astor wrote his check with which we started this chapter, exhibits extraordinary variation and complexity in code choice. Roman numerals, Western numerals, and English numeral words are not in competition with one another, but rather complement one another in producing a multimodal text rich with numerical information.

Some of this variation in parallel modalities may have served to clarify or prevent alteration (VIII vs. *Eight*, 13 vs. *Thirteen*), but clearly this is not a full explanation. Rather, viewing this text and others as complex fusions of different modalities—not employed haphazardly, but rather chosen by those who had the agency to select them—and reflecting on why they made the choices they did, allows a richer understanding of all sorts of numerical material around us. For instance, the use of Roman numerals for the amount in dollars—a purely American currency—versus Western numerals (at the bottom, center) for the amount in pounds, takes on a new meaning when we see that this bill was produced just a few months after the signing

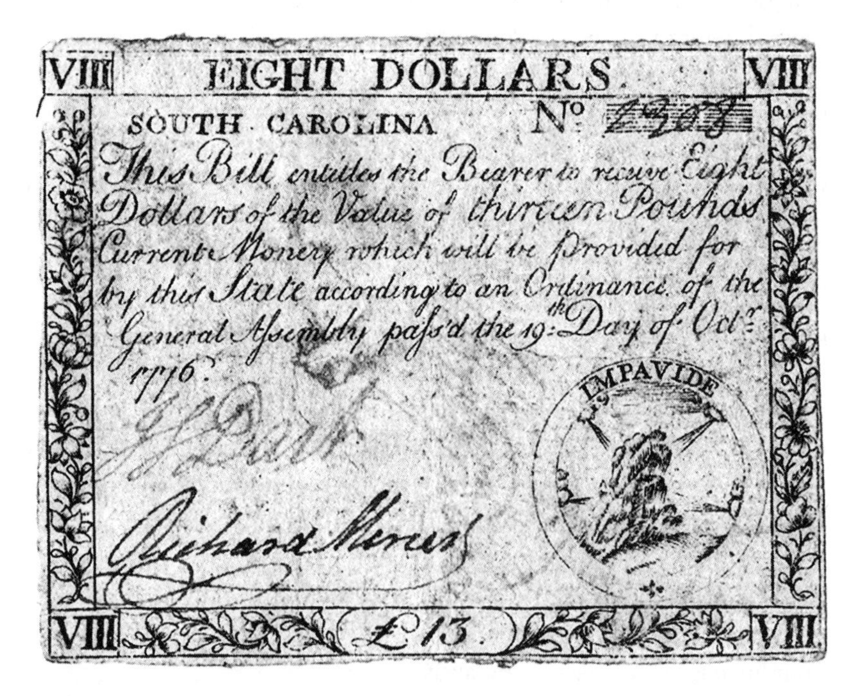

Figure 6.15
South Carolina eight-dollar bill, 1776

of the Declaration of Independence. But the number 2308 handwritten on the bill as its serial number is, understandably, in Western numerals—quite a convenience when some poor writer had to annotate each bill in turn.

Conclusion

Sociolinguists often talk about *code switching* or *code mixing*, both in discussions of verbal speech and in analyses of texts, because people do not always speak only in one language at any given time. Some sociolinguists use code switching to refer to switching across different phrases or sentences, and code mixing to refer to switching languages within an utterance or phrase. Under this model, parallel modalities most closely resemble code switching, while blended and hybrid modalities are closer to code mixing. But more important than a typology of different kinds of language mixing as they relate to notations is identifying the cognitive, social, and textual explanations for the phenomenon. Multilingual people, sometimes for reasons of imperfect fluency but most often for reasons of style, emphasis, or identity marking, engage in code mixing and code switching regularly (Myers-Scotton 1993). We might want to know what we can learn from these sorts of instances—how choices are made and deployed, and why.

Those of us who study writing and literacy, similarly, are used to thinking about bilingual and trilingual texts. Most schoolchildren still learn about the decipherment of Egyptian hieroglyphs using the example of the Rosetta Stone written in two languages (Greek and Egyptian) and three scripts (Greek, demotic, and hieroglyphic), and the artifact has come to serve as a materialized metaphor for translation and decipherment in general. There is also a vast and growing literature on the multilingualism of individuals and groups in antiquity, with code choice always being present as writers used different languages promiscuously across a range of genres (Crisostomo 2015; McDonald 2015). Who writes and for whom? What code(s) do they use and why? These questions motivate many authors in Near Eastern studies, classics, and numerous other historical disciplines. In contrast, number systems are seen almost as needing no decipherment, needing no bilingual texts, needing no explanation—in other words, not of much interest except for mathematicians. After all, even where there are undeciphered scripts, we can often read the numerals. The Maya numerals were deciphered almost a century before the phonetic aspects of the Maya script were revealed. The Minoan Linear A numerals were fully understood at the time of Sir Arthur Evans (1909), and are still the only part of that still-undeciphered script about which there is some agreement.

We can see now that this perspective is a mistake. There is almost always some variation, and thus some choice, as to how to write a number, lexically, graphically, or using some combination of principles. Even a text written in a single language, and intended for a narrow audience, affords the writer opportunities to select some combination of expressions that suits a set of needs and goals. Very often, multiple and sometimes competing principles and interests will be at play, making the explanation of particular choices all the more fluid. Some of the explanations for these choices include, but are not limited to:

- *Multilingualism*: Because not every reader can be assumed to be fluent in a particular language or script, numerical notation serves as a translinguistic bridge.
- *Clarity*: Because numerical notation is visually salient within a text, adding numerals provides clarity and makes the task of identifying and reading numbers in a text easier.
- *Security*: Using two distinct representations of the same number provides security, in case of a scribal error or alteration, that the number represented is that intended.
- *Familiarity*: Rare numerical notations—very new or obsolescent, e.g.—may need explication or linguistic redundancy.
- *Conciseness*: Mixing modalities can in some cases be shorter than either modality used alone (12 billion vs. 12,000,000,000 vs. twelve billion).
- *Aesthetics*: Code choice involves aesthetic as well as practical considerations: prestige, modernity, playfulness.[6]
- *Audience*: Numerals are chosen that are likely to be understandable or valuable to particular imagined readers of the text, as opposed to others.

Numeration is thus not a mere appendage to scripts. It displays remarkable variation in form and function in practically every literate society. It is not just "math stuff"—in fact, most of it isn't for math at all, but simply solves the problem: how to write a number? Because there are multiple ways to write numbers in almost every tradition, numeration is a rich source of insight into how premodern writers thought about problems of representation. It also helps us think about reading and readership in environments where familiarity with number symbols was almost surely more widespread than full literacy.

Once we can recognize the complex set of factors that motivate these choices, we can return to the broader picture, and to larger historical time scales, as we look for patterns and processes that apply on a comparative basis across multiple millennia. The micro-scale choices made at the level of individual writers come back to inform how we think about much longer-term patterns of change across deep time.

7 / VII To infinity and beyond?

To conclude, I return to reckon once more with the central questions that motivate this book. Some of these are specific to number systems and numerals. Some of them are far broader. And by "reckon," I do mean both that I hope to think about them and come to some sort of judgment—with the proviso that reckoning is always an ongoing process, never complete.

Is your number system weird?

On the surface, this question may seem weird, and you may be thinking that its author is weird too. But I mean it seriously. We have these numerals, 0123456789, and we know how to combine them using place value. We begin our thinking lives, as infants, surrounded by numerals—if you grew up in any Western country you were surely exposed, almost from infancy, to brightly colored letters and numbers to look at, chew on, and think about.[1] One result of this exposure is that, largely unconsciously, you came to accept that your numerals are normal and natural. This exposure is also the root of human ethnocentrism, that feeling we all experience where things to which we have been exposed habitually are easy to take for granted, while unfamiliar experiences produce reactions ranging from mystified wonder to disgust. It's not good, but ethnocentrism almost surely is inevitable.

So my first reaction to the question of whether the Western numerals are weird might be that every number system is weird, viewed from the outside. This is a form of particularism, the notion that ideas and practices are unique configurations existing at particular points in time and space. Particularism is widespread in the social sciences and humanities, and for good reason: there are lots of things that are local and unique but are nonetheless highly worthy of our attention. There is a lot of emphasis in popular and everyday discourse about science on generalizability, in a way that suggests that we ought, always, to be looking for the most generalizable facts about the world; and by that logic, only universals ought to be of interest. But generalizability is always about generalization *to some set of cases or examples*—we do not presume, for instance, that biologists who study tapeworms ought to be able to generalize their findings to earthworms or manatees or orchids. There's nothing wrong with studying the particulars of something.

But, you might rightly object, sometimes we really do want to compare like with like, apples with apples, tapeworms with tapeworms, numbers with numbers. In chapter 1, I outlined some of the things in which we might be interested, including identifying rarities in numerical systems—things that only happen very rarely, but nonetheless that do happen. For instance, the vast majority of the world's languages have a base of ten, with twenty also being popular. A sample of 196 languages in the *World Atlas of Language Structures* revealed that 125 languages are decimal, 20 vigesimal (base-20),[2] 22 a mixed decimal-vigesimal, 20 restricted (having few enough number words to need no base), four having a system using body parts as numerals, and five having some other base (Comrie and Gil 2005). Decimal systems are found on every inhabited continent, across dozens of language families. This simply cannot be a coincidence. We would rightly conclude that, while "normal" is too loaded a word, "common" certainly applies to decimal number word systems. In contrast, within that sample, only one living language, Ekari (also known as Kapauku), spoken in Papua New Guinea, has a base-60 (sexagesimal) system.[3] We're fairly used to thinking sexagesimally because our time system has base-60 elements, but we certainly don't have any special number words or symbols for 60 or its powers (3,600). In fact, the only other language to have a sexagesimal structure that I'm familiar with is Sumerian, last spoken millennia ago in Mesopotamia, and whose structure influenced the large round numbers on the Weld-Blundell prism discussed in chapter 2 (Powell 1972). Do we conclude, as some have done, that Ekari numerals were borrowed from Sumerian (Pospisil and Price 1966)?[4] Barring any historical or other linguistic evidence (of which there is none), the only way we could do that is to show that base-60 numeration is so weird that it could only have happened once. But how would we know that? If it happened once, why could it not have happened twice?

Because our knowledge of premodern languages is so limited, because written records are extraordinarily partial and skewed toward elites, and because historical linguistics depends on modern languages to reconstruct past ones, we have no idea how many times sexagesimal numeration was invented, simply on the evidence. We need some body of theory to help us explain why some notations are common and others are rare, before we could possibly answer the question of what is weird and what is not.

We actually know very little about what kind of sample the world's present 7,000 or so number word systems is, out of all the linguistic variability that has ever existed in the domain of number. Most of that is permanently lost to us, and even the parts that we can reconstruct using historical linguistics depend on the survival of elements of past languages in their modern descendants. We are on firmer ground—though still not completely secure—with regard to the world's numerical notations. They're visual and relatively permanent, and we can be relatively sure that there weren't any base-structured notations (the five basic types I have outlined) prior to the advent of literacy around 5,500 years ago. But if the Cherokee numerals I described in chapter 5 can come so close to slipping out of our knowledge, though invented less than two hundred years ago, there's every reason to think that there are many more for which evidence does not survive—or has yet to be found.

Thus, the answer to whether a number system is weird or not depends partly on the accumulated empirical evidence of number systems past and present, but also on some set of theoretical principles about what sorts of processes we imagine led to their development. Ten years ago, I would have sworn up and down that induction was the best approach in the historical disciplines—that while we can never aspire to absolute truth, when in doubt we should stick to what we can know directly. But a purely inductive approach only handles well the sort of data that we have at hand. I don't think all is lost, but we need reasons to hypothesize that our comparisons are good ones. There are good theoretical reasons to suppose that some things, like subitizing, are panhuman, and that some structures, like decimal notation, were cross-culturally just as common today as in the past. And, while it's difficult to reason about things for which there is no direct evidence, if all known actual numerical notations fit into the five basic and common structural types (even if we can imagine numerical notations with other structures, and even if there were *really* weird ones that didn't survive), one reason for it may be that the others weren't "good to think with"—that they violated some general principle or norm that human brains prefer. But this brings us to a second issue, which is the prospect, frightening to some, joyous to others, that comparison across time scales may always fail to account for the particular weirdness of modernity.

Is the past like the present?

Or, rather, should I ask: Was the past like the present?

This may seem like a bit of linguistic chicanery (fair enough), but I also want to draw our attention to something that historians, archaeologists, historical linguists, and for that matter geologists and cosmologists are trained to be aware of—that we can only study the past in the present, through its traces. But at the same time, to deny the past its own reality is to commit a sort of radical present-centrism, which we would never tolerate in other forms of ethnocentrism. No one doubts that there were, once, people, 100 or 500 or 5,000 or 10,000 years ago, who lived lives and did many of the things that we do. One need not be a radical realist to commit at least that far.

So, better then to ask (though more clumsily): to what extent can we establish whether the range of variation in aspects of the human condition today is similar to, or different from, the total range of variation at all times and places? We can constrain ourselves a bit by restricting ourselves to *Homo sapiens sapiens*, so that we do not have to deal with many of the problems of our evolutionary history. This would be an arbitrary convenience for an analysis that aims to look at human cognition.

Donald Brown's remarkable text *Human Universals* is one of the most underappreciated works of anthropological theory of the past few decades (Brown 1991). Universalism is not in vogue in anthropology, partly for understandable reasons and partly due to contingent and ephemeral discipline-specific biases. The central premise of the work is that one of the more important things that anthropologists can do to theorize the human condition is to pay attention to invariant aspects of human existence, and that these are more numerous than we might think. As I discussed briefly in chapter 1, Brown is keenly aware that universals are time-situated. He notes that some universals are clearly *former universals*: things that used to be true of all humans, but are no longer so. Prior to the development of agriculture (which the archaeological record tells us happened independently in many different parts of the world), all humans were forager-hunters, but today almost no one is. This is not a triviality—people 15,000 years ago were, to the best of our ability to discern, biologically and cognitively very similar to us, and if something (foraging-hunting) that held for millennia can change, we ought to be very careful asserting that something must be true of all humans, forever and always. Conversely, Brown identifies *new universals*: things that were formerly rare or at least nonuniversal, but are now universal. The domesticated dog is one proposed new universal Brown gives us, but we can think of other potential ones in an age of rapid modernization and globalization. The point is clear, then: any theoretical framework that aims to assert broad truths about the human condition must be aware of the degree to which its claims are time-specific.

Henrich, Heine, and Norenzayan's (2010) article "The Weirdest People in the World?" draws on a host of insights from cross-cultural anthropological and psychological research to argue, correctly, that too many of the findings of behavioral sciences are limited by the fact that the samples from which they generalize are excessively Western, Educated, Industrial, Rich, and Democratic (i.e., WEIRD). Just as the fictional Nacirema seem hopelessly weird to anthropology undergraduates until they realize that their customs are actually those of Americans (Miner 1956), most of the academy is populated by WEIRD people who just haven't realized how weird they are. To their formulation, though, we need to add a sixth letter, M—at the risk of losing a beautiful acronym—because the populations studied by behavioral scientists are almost universally Modern. If the populations today are not like the populations of the past, then any generalizations based on ethnographic data, or psychological experiment, or survey alone will be problematic if extended beyond their time. For anthropologists, not only is the ethnographic record a biased sample of all societies that have ever existed, it is biased to an unknown degree and in unknown ways. To put it another way, we do not know the size of the statistical universe of societies of which the ethnographic record is a sample. Without denying that WEIRD societies are profoundly weird, past societies may be equally weird, but in ways not currently apparent to us.

Thus, for any topic, for any domain of experience, across any branch of anthropology (or any other human science), to build theory is inevitably to choose between a present-centered particularism (which is fine, as long as we recognize it for what it is) and a temporally broad, cross-cultural comparativism that seeks to expand our sample of human variation. So, for instance, an anthropology of the state that takes its start from the work of Eric Wolf (1982) and that regards Western globalization as the principal subject of study is very different from one that situates contemporary state theory in a temporally deep framework, such as recent work by James Scott (2017) or David Graeber and Marshall Sahlins (2017). Wolf's work starts with the presumption that the relevant history for understanding the modern state is the development of global systems of wealth extraction and power inequalities starting around 1400 or so—i.e., capitalism and its predecessors, centered in Western thought. It is extraordinarily important for counteracting, systematically and empirically, the notion that the ethnographic record reflects pristine peoples untouched by historical and social forces. But in doing so, it also separates us from the deeper past. In contrast, the more recent comparative work on the state takes the view that understanding these inequalities requires a consideration of the millennia of deep, pervasive inequality that preceded them, while noting equality and antihierarchy where they existed, and insisting that inequality is not inevitable. There's lots to take from both these perspectives, and theoretically they share a common substrate of historical materialism, but they entail very different views about the value of studying large swaths of humanity.

One could object at this point that for some topics, there's good reason to think that the past and the present might be really different. So let's take that idea seriously, and think about two different possibilities.

Hypothesis 1: The present is really dissimilar to the past. Perhaps the present system of globalization, industrialization, mass media, etc. is radically different from anything that has existed previously. In so many ways, this hypothesis is so obvious that it's taken for granted in many circles. It can't be limitless difference (we're all human, after all), but it sure could be big enough to warrant treating the past couple of centuries as incomparable to earlier times. Now, if you're an ethnographer, depending on whether you're an optimist or a pessimist, you can interpret that in two ways. You might say, "Well, as an ethnographer, that means I don't need to worry about anything that happened in the past—basically, archaeology is a radically different subject matter because it deals mainly with periods that are completely unlike how anyone lives today." On the other hand, you might say, "Uh-oh ... if the past is really that different, then I need to be aware that I'm just describing a small sliver of humanity, not just in space but in time." While the first answer goes against the idea of a broad, temporally deep approach, the second embraces it; but they rely on the same insight.

But the idea that the past and present are dissimilar is not the only possibility.

Hypothesis 2: The past and the present are not so different after all, but the nature of the data and methodologies used to learn about them are. The problem is then an epistemological one, not an ontological one. Perhaps the nature of the data in the ethnographic record is such that comparable phenomena existed in the past but simply haven't survived, or are not amenable to archaeological or historical analysis. Ethnographic data are collected at particular moments, and ethnography records data at a micro scale compared to what archaeology records. You get different sorts of things than you get if you're an archaeologist, where even a span of 50 or 100 years—whole generations—is considered brief. This is related to the problem that Martin Wobst called the "tyranny of the ethnographic record" over forty years ago (Wobst 1978). In particular, Wobst was challenging the notion that we should primarily use models and methods based on ethnographic data to explain archaeological evidence for hunter-forager behavior. It wasn't that he thought the past and present were different—it's that he thought that archaeology and ethnology were different. But again, there are two possible approaches. Our hypothetical ethnographer could say, "Method largely determines the questions we ask and the answers we get. As an ethnographer, what I get is really going to be incommensurable with what my archaeologist buddy gets, even if we're working in the same region. The archaeology of hunter-foragers is thus irrelevant to what I do." On the other hand, another ethnographer might say, "Uh-oh ... if I'm honestly interested in getting past my methodological perspective, I'd better find some other complementary perspectives to work alongside mine." Again, two diametrically opposed answers to the same observation.

I am sympathetic to Wobst's argument insofar as it recognizes that ethnographic analogies may misconstrue prehistoric data, in particular because the methods and time scales of the two subfields differ, and insofar as it attempts to rebalance the weighting of ethnographic and archaeological data. But, to the degree that Wobst's arguments have served as a rationale for archaeologists and ethnographers to ignore one another on the assumption that their datasets are incompatible, they are fundamentally misapplied. I am arguing here for a much closer coordination of archaeological and ethnographic data, not despite the fact that they are different, but because they are different, and because we need to explain that difference if we are to have any success in understanding the range of constraints on social formations. I am far more sympathetic to the call for cross-cultural comparisons of archaeological as well as ethnographic data of the sort that the archaeologist Peter Peregrine (2001, 2004) has advocated, adding a diachronic dimension to a previously synchronic enterprise.

The archaeologist André Costopoulos goes still further, and argues that our total knowledge of the universe of possible configurations of human societies (including ethnographic, historic, and prehistoric data) is "a sample of the universe of possibilities whose relationship to that universe is unknown to us. The extent and composition of the universe are completely unknown. Even for the tiny portion of the space that is documented, there are significant disagreements about the number of objects and the ways in which we can circumscribe them" (Gray and Costopoulos 2006: 151). This is a serious challenge to all comparative social scientific research. The answer, again, is that comparative research cannot be conducted in a theoretical vacuum. Cross-cultural research questions acquire validity as part of frameworks of research, rather than as a random inductive search for patterning. For numeral systems, an analysis informed by what we know about numerical cognition is justified as the foundation of the search for patterns in the world's known number systems, past and present.

Why is there no medieval anthropology?

But the problem is still deeper, because even if data from prehistoric archaeology were used systematically to complement our knowledge of the ethnographic record, anthropology has never asserted as its purview, or, more cynically, actively removed from its purview, a wide swath of societies known principally through historical evidence but not mainly through archaeology, and in particular, that constellation of societies falling under the somewhat inapt but hardy label "medieval." The study of the so-called ancient civilizations such as Egypt and Mesopotamia is disciplinarily diffuse but includes numerous anthropological archaeologists. There are dozens of archaeologists trained jointly in anthropology and classics departments, and/or who teach interdisciplinarily in both fields. Hundreds of classical archaeology students every year get their archaeological training primarily in anthropology departments. Similarly, there are hundreds of early modernists in anthropology: people who focus on Spanish colonialism in the New World, for instance, or world systems theorists, or people interested in Atlantic World / diasporic studies, or really most of the folks who would describe themselves as ethnohistorians. The Middle Ages, understood roughly as the period from around 500 to 1500 CE, is the only period to be so deeply underrepresented in anthropological theory and empirical research.

One problem that medieval historians face is that "the medieval" is a nebulous object, subject to both scholarly and popular imposed definitions that satisfy no one. "Medieval India," "medieval Japan," "medieval Islam," etc. all refer to very different social configurations and chronological periods. If there were a purely chronological definition of "medieval," then one would really need to include New World societies as well. Hardly anyone talks about the "medieval Maya" in the way that people seem very happy to talk about "medieval Japan." As of October 2018, the former phrase has only 8 results in Google Scholar, while the latter has 8,800! There is no methodological justification for including "medieval" Mesoamerica but excluding medieval Islam from anthropological investigation—this is strictly a matter of arbitrary disciplinary divides that, at worst, can be construed as racialized. Fortunately, medieval scholars have recently begun to construe a "global Middle Ages" that might include the Americas, or at least to think beyond the traditional bounds of Europe and the Mediterranean to consider broad patterns of culture contact (Jervis 2017; Keene 2019). But anthropologists have not had much to say on the matter—we have not been part of this discussion.

We have very little conception of what a medieval anthropology would look like. Only a handful of anthropologists have given serious consideration to Old World societies between 500 and 1500 CE (e.g., Goody 1983; Hastrup 1985; Hodgen 1952; Macfarlane 1978). Raoul Naroll's largely unheeded call for "holohistorical" work, which would insert historical data into broader cross-cultural analyses, could help fill that gap (Naroll et al. 1971; Naroll, Bullough, and Naroll 1974), but only if anthropologists acknowledge that anthropology cannot be a fully comparative discipline without this data. But the medieval is almost entirely out of our grasp. I recall being amazed when, as a graduate student working on historical issues in anthropology, I first encountered A. L. Kroeber's once-classic textbook *Anthropology* and realized it had a whole chapter on the invention and diffusion of the zero in medieval India, the Middle East, and Europe (Kroeber 1923). This sort of subject matter is foreign to contemporary anthropology. This is an odd gap, to say the least, for a discipline that purports to be a holistic comparative study of human behavior. In many small (and not-so-small) history departments, medievalists get the dubious honor of teaching Western civilization courses that start with Sumer (in which anthropological archaeology has numerous specialists) and end with the twentieth century (of which the vast majority of cultural anthropologists have some knowledge). It's not that I think that I, or any other anthropologist, would do a better job than a medievalist would of teaching such a course, nor would I want to do so. But if I were going to construct a "world survey" anthropology course, it would be very challenging to come up with relevant material written by anthropologists or anthropologically trained archaeologists that focuses on the millennium of history in which medievalists specialize. But I can't think of any valid conceptual or methodological reason to exclude the medieval from the anthropological.

To be fair, there are some notable exceptions. The social anthropologist Jack Goody, whose work on literacy I have already discussed and is an important precursor to this work, treated the medieval as a subject for serious study, across domains as disparate as literacy, food, and marriage (Goody 1977, 1982, 1983). Alan Macfarlane, with doctorates both in history and anthropology, writes on witchcraft, capitalism, individualism, and marriage through a comparativist anthropological and historical perspective in which the medieval receives heavy attention (Macfarlane 1970, 1978). Historical anthropology in Iceland is similarly weighted toward the medieval, partly because of the unique population history of the island (Hastrup 1985; Byock 1990; Durrenberger 1992). In Americanist anthropology, there are two audacious, if ultimately peripheral, efforts of note that incorporate the medieval. Margaret Hodgen, whose career was sadly derailed from what it might have been by her unwillingness to participate in anticommunist loyalty oaths in the 1950s, wrote a series of important papers on the subject of cultural diffusion and innovation using medieval and early modern data (Hodgen 1945, 1952, 1974). In his later years, Gordon Hewes, a true four-field holistic anthropologist and a student of A. L. Kroeber, amassed thousands of pages of material for a comparative history of the seventh century CE, of which only small and programmatic fragments were ever published (Hewes 1981, 1995). I mention these last two not because they are important, but rather because they have had such marginal influence within anthropology.

Yet we must be wary not to simply use "medieval" as an indicator of stage without consideration of chronology. In his massive, almost primordial volume of anthropology, *Primitive Culture*, E. B. Tylor wrote, "Little respect need be had in such comparisons for date in history or for place on the map; the ancient Swiss lake-dweller may be set beside the medieval Aztec, and the Ojibwa of North America beside the Zulu of South Africa" (1871: 6). At a glance Tylor is allowing for a broad anthropology including historical societies, but only at the cost of locking past peoples into stages, only comparable to others of the same type. This dehumanizing comparativism is far worse than analyses that look only at the present, because data are ripped out of any context that might be relevant to understanding them, and retrofitted to a rigid hierarchy of social standing. These problems of unilinear cultural evolutionary theories are so well known in modern anthropology as to need no further exposition.

The archaeologist Shannon Dawdy, in an essential article on modernity, ideologies of the past and present, and the way in which subjects of inquiry are included and excluded from anthropological knowledge, confronts a similar problem in her account of "clockpunk anthropology"—one that treats chronology, not by ignoring it or turning it into stages, but by recognizing that past and present concerns are not so different (Dawdy 2010). Dawdy's problem, as a historical archaeologist, is a different one from mine, in that her subject matter, the analysis of ruins from relatively recent American cities, is separated by disciplinary disjunctures from its counterparts in antiquity. She rightly bemoans the neglect of "modern" ruins simply because some archaeologists regard them as insufficiently archaeological to be of interest. Dawdy's work shares with my argument a broad commitment to comparativism as a framework for breaking down rigid periodizations and disciplinary silos. Challenging what she sees as a rigid divide between modernity and antiquity, she argues that archaeologists who work on prehistoric and historic periods have much more in common than either of them normally allow. We need to be able to ask better questions, such as:

> Are the differences between ancient and modern cities simply those of scale and tempo, or are they truly of kind? Are grid patterns and secular subjects such whole new inventions? Or totalitarian architecture and panoptic public spaces? What would ancient Greek and Roman urban sites reveal about our own spaces? Or

> those of Tenochtitlan and Teotihuacan? Most archaeologists of antiquity decline to consider the possibility of such modern phenomena as racialization, capital accumulation, or terrorism in the past—to look for such things in antiquity is not only anachronistic but also offensive. The past is not supposed to share these dystopian aspects of our present and recent past. The deep past is, for many, a utopian refuge. (Dawdy 2010: 364)

These are important questions indeed, and important problems! From my perspective, there are similar omissions in the medieval period, and in much of antiquity that is considered "historical," because anthropologists, archaeologists, historians, philologists, and others all seem to have agreed upon a division of labor whose effects on our ability to ask good comparative questions—about number systems or anything else—are pernicious.

I have spent a lot of time in this book talking about the Roman numerals, partly in antiquity but chiefly in the late medieval and early modern period, just when they were beginning to be seriously challenged for dominance in Europe by the newfangled (at least to Europeans at the time) Western numerals. There has never been a full-length English monograph on the Roman numerals, which is surprising given their antiquity, their widespread use, and their continued cultural importance. Nor do I claim that this book serves this purpose. To write a history of the Roman numerals would be to grant that they are *sui generis*, to reaffirm that they are a subject of antiquity and the medieval, and perhaps even to undermine my own credibility to comment on these matters (as I am neither a classicist nor a medievalist by training). Instead, by framing the decline of the Roman numerals as merely the most prominent case among a comparative set of notations, ancient, medieval, and modern, and governed by the same sorts of constraints and cognitive factors, I hope to historicize these phenomena, but only to the degree they warrant. And we do not know that degree in advance—it can only be established through investigation.

What is the future of numeral systems?

There is a strange paradox in the progressivist framework under which much of Western thinking about technology operates. On the one hand, there is, for many of us, a perception that time is speeding up, that social change is happening more rapidly now than ever before, and perhaps even that the rate of acceleration is itself accelerating. Partly this is surely a cognitive bias derived from a combination of a certain "good old days" brand of traditionalist narrative in the media, techno-utopian futurist notions of progress, and the fact that at the scale of the life course, one's formative years seem to pass more slowly than one's later life. This logic leads to some varieties of transhumanism, a viewpoint that contends that, in the near to medium future, technology's interface with the human body and specifically the human brain will accelerate beyond the point where we can be reasonably considered human, and ultimately, through superhuman intelligence, will lead to a technological singularity the scope of which we present humans can hardly fathom (Kurzweil 2005; Bostrom 2014). Under this frame of reference, we have absolutely no reason to expect that the numeration systems of today will be anything but laughable to humans perhaps only a century in the future. Futurists may not be thinking, specifically, about the "Numerals of the Future," but as numerals are the product of modern human cognition, we can readily see that the numerals of today may not be suitable for the posthuman tomorrow.

On the other hand, there is an almost perverse certainty, in many domains of existence, including number systems, that the people of today have come up with the One True Answer to many central problems, and that essentially no further progress can ever be made in them. Much mockery was made of the political scientist Francis Fukuyama when he published *The End of History and the Last Man* (1992), signaling his belief (which he now largely rejects) that with the fall of communism, liberal democratic capitalism was the final social configuration of human societies. The notion that the next millennium, or five, or ten, of human existence will have a single mode of production and a single form of political decision making is nonetheless still widespread. Partly this is just a failure of our imagination (although, for the record, many such alternative models are available, across political spectrums and philosophical frameworks, if you care to look). The idea that we are at the "end of history" of numeration is also widespread. As the cognitive neuroscientist Stanislas Dehaene remarks:

> If the evolution of written numeration converges, it is mainly because place-value coding is the best available notation. So many of its characteristics can be praised: its compactness, the few symbols it requires, the ease with which it can be learned, the speed with which it can be read or written, the simplicity of the calculation algorithms it supports. All justify its universal adoption. Indeed, it is hard to see what new invention could ever improve on it. (Dehaene 1997: 101)

All right, you might respond: what might such a new "killer" number system look like that might replace the Western numerals? And I might demur that I'm not a futurologist, simply noting that it's unlikely that ten thousand years from now, our descendants will still be using ten digits with place value. While there are notations, like the Egyptian hieroglyphs or the Roman numerals, that survive for a millennium or two, even these undergo tremendous change over their period of active use—so why would we expect that change to have stopped? My job is not to make up new systems, just to describe them. But I can see that this response, although accurate, is not the best possible one, because several numerical notations explicitly designed to be better than Western numerals (for some value of "better," for some purpose) are already on offer.

First, there are systems developed by groups like the Dozenal Society of America. Founded in 1938, it was originally called the Duodecimal Society of America for some decades until it was decided that this was too decimal-centric. The Society advocates for base-12 arithmetic in place of base-10, principally because 12 is a highly composite number with many factors (1, 2, 3, 4, 6, and 12), making work with multiplication and division more convenient (Andrews 1944; Brost 1989). For numerals, converting to base 12 would simply be a matter of adding single digits for 10 and 11 (the Dozenal Society prefers ↊ and ↋). But that's a fairly minor change in the big scheme of things—for most of us, if we ever think of "alternative number systems," choosing a different base is a pretty obvious variant. Dozens (or at the very least, *tens*) of science fiction writers have created weird number systems for alien or future human societies (Pohl 1966), and most of them just use a different base along with ciphered-positional notation.[5] You may be familiar with hexadecimal numerals used in computing, whereby the letters A through F represent 10 through 15, so that 1A = 1 × 16 + 10, or 26, as an accommodation between the binary nature of electronics and the constraints on conciseness that make actual binary numbers too long for the human eye and mind. But let's get weirder.

In 1905, the mathematician and tidal scientist Rollin A. Harris (1863–1918) published a now nearly forgotten article entitled "Numerals for Simplifying Addition" in the *American Mathematical Monthly* (Harris 1905). Harris begins by noting that many of the number systems of antiquity are self-evident in meaning, while the digits 0 through 9 are obscure in their meaning, and simply have to be learned. In terms of the typology I outlined in chapter 1 and elsewhere, the systems he's interested in are cumulative-additive—using repeated signs that are added together, while Western (which he calls "ordinary") numerals are ciphered-positional. Harris proposes, then, to reform our digits, keeping the decimal, place value system intact but reshaping the digits themselves using combinations of vertical strokes for 1, horizontal strokes for 2, and circles for 5, partially along the model of the Syriac numerals (figure 7.1). Because these signs are additive—the structure of each individual sign combines fives, twos, and

Ordinary	Proposed	Syriac	Roman Rep. Period	Palmyrene	Phoenician	Hieroglyphic
0	[illegible]					
1	/	/	I	/	I	I
2	[illegible]	[illegible]	II	//	II	II
3	[illegible]	[illegible]	III	///	III	[illegible]
4	[illegible]	[illegible]	IIII	////	[illegible]	IIII
5	o	[illegible]	V	[illegible]	[illegible]	[illegible]
6	b	[illegible]	VI	[illegible]	[illegible]	[illegible]
7	[illegible]	[illegible]	VII	[illegible]	[illegible]	[illegible]
8	[illegible]	[illegible]	VIII	[illegible]	[illegible]	[illegible]
9	[illegible]	[illegible]	VIIII	[illegible]	[illegible]	[illegible]
10	[illegible]	[illegible]	X	[illegible]	[illegible]	[illegible]

Figure 7.1
Proposed "simplified" numerals along with several ancient analogues (Harris 1905: 66)

ones as needed to reach any number up to nine—they can be read just as a Roman VIII can be read as a five followed by three ones.

By adopting these new signs, Harris argues, arithmetical errors will be reduced, the numerals will be learned more quickly and with less effort, and children will more rapidly learn addition. A modern cognitive scientist could argue (although we don't have any experimental evidence) that they offload more of the reader's and writer's cognitive load to external representations, and thus require less internal, mental cognitive work (see, e.g., Zhang and Norman 1995). A semiotician might note that these signs, although not exactly iconic, are more highly motivated than our present, opaque ones. They preserve all of the advantages of the Western numerals and add to them a new additive clarity in the sign forms. But of course, their adoption never happened—in fact, Harris's work does not seem to have been cited by anyone (except me) in the century or more since its publication. We can look at this system and see that it is, in some sense, superior to Western numerals, retaining all their advantages while improving on others—and, at the same time, we can see that it would have been very difficult indeed for this early twentieth-century innovation to be adopted and replace the Western numerals. As clever and innovative as it is, Harris's system seems to be a solution to a problem that no one was facing.

In 1947, the amateur mathematician James E. Foster showed that it was possible to have a number system that was infinite, concise, but that had no zero symbol—rather, it used a T for 10. So, for instance, counting up from 98, we have 99, then 9T (9 tens + ten), T1 (ten tens + one), T2 (ten tens + two), and onward up to 1,000, which is 99T (nine hundreds, nine tens, + ten) (table 7.1). This system is, at first glance, very strange, but it is totally unambiguous, decimal, positional, and infinitely extendable. Most of its numerals are identical to Western numerals, but numbers that we would represent with zeroes look very different. It is also more concise than Western numerals—compare 100 to 9T or 1,000 to 99T.

You might object, at this point, that 9T is completely opaque to English speakers who are used to 100. But then again, who's to say that the user of this system would be an English speaker? In creating this system, Foster was no reformer or technocrat—he was also the author of "Don't Call It Science" (1953) and "Mathematics Need Not Be Practical" (1956), and thus an advocate, following G. H. Hardy's classic *A Mathematician's Apology* (1940), of a humanistic, deeply impractical mathematics. Unlike Harris, his goal

Table 7.1
Zeroless numerals from 1 to 120 (11T) (after Foster 1947)

1	2	3	4	5	6	7	8	9	**T**
11	12	13	14	15	16	17	18	19	**1T**
21	22	23	24	25	26	27	28	29	**2T**
31	32	33	34	35	36	37	38	39	**3T**
41	42	43	44	45	46	47	48	49	**4T**
51	52	53	54	55	56	57	58	59	**5T**
61	62	63	64	65	66	67	68	69	**6T**
71	72	73	74	75	76	77	78	79	**7T**
81	82	83	84	85	86	87	88	89	**8T**
91	92	93	94	95	96	97	98	99	**9T**
T1	**T2**	**T3**	**T4**	**T5**	**T6**	**T7**	**T8**	**T9**	**TT**
111	112	113	114	115	116	117	118	119	**11T**

was not to advocate that such a system should be adopted, but to show that it was infinitely extendable, had a single representation for each number, and that it was easily manipulable as a mathematical system. But the advantage of conciseness is there nonetheless, if that's your interest.

As it turns out, while Foster was the first to invent such a system, his creation has been (apparently) independently invented several times since. The mathematician Raymond Smullyan, in his classic *Theory of Formal Systems* (1961), called such systems *lexicographically* arranged—their order is from shortest to longest and, within strings of a given length, "alphabetically" from lowest to highest. Robert Forslund (1995), seemingly unaware of Foster's precedent, redevelops it and then argues that premodern inscriptions and texts using such systems might have been misidentified as errors, calling for "archaeologists specializing in the interpretation of these ancient documents to examine this usage." Fortunately for us, there do not appear to be such longstanding errors of interpretation, but Forslund's general point is correct: even if this system was never used, there is no reason why it could not have been invented.[6] After all, he invented it unaware of Foster's creation!

The mathematician Vincenzo Manca (2015), building on Smullyan's lexicographic ordering, invented a system identical to Foster's (using X instead of T for 10, but otherwise the same) and noted a further benefit. These systems are known as *bijective* numeration systems—unlike Western numerals, where one can (as in identification numbers) insert leading zeroes (so that 1, 01, 001, 0001, etc., all represent 1), this system has a unique identifier for each number, so any set of strings can be readily ordered. There are no leading zeroes, since there are no zeroes at all. Manca's use for this sort of system was to produce an ordering of DNA sequences viewing the four letters A, C, G, T representing the bases as a sort of base-4 numeration system. Unlike Foster, who started with a concept without any notion of practical utility, Manca started with a practical problem and then hit on the same solution.

I'm not mentioning this system because I think it is the numeral system of the future, or because it's so wonderful that we should adopt it. Frankly, my goal is to flummox you. Staring at this weird system, bewildered, is about as close as we can get to the mindset of those Indian, Middle Eastern, or Western European mathematicians upon first encountering decimal, positional notation between the sixth and eleventh centuries. How strange it seemed to them at first glance, this idea of place value with the zero. It was a subject of bafflement. It took centuries to be fully accepted, and it needed to be explained in great detail to new users. The zero, in particular, was especially confusing, and was described by one late twelfth-century author as a "chimaera"—a monstrous digit, a number and yet not a number (Burnett 2002). The philosopher of science Helen De Cruz argues that some numerical concepts, like the zero, may be difficult to accept or adopt because they are deeply counterintuitive, although, once accepted, their weirdness may be appealing (De Cruz 2006: 317). That feeling you probably have right now, that this monstrosity with Ts for 10s couldn't possibly be workable, possibly mixed with a protective paternalism for the beloved Western numerals, is what often happens when novelties get introduced. It is partly because their unfamiliarity requires effort—the QWERTY effect discussed in chapter 4. But it is also because the functions for which this new system would be useful are not those most relevant to our present concerns.

But perhaps you want a system that actually has a demonstrable, current use. As I argued in chapter 1, there is a gap between the imaginable and the attested—those things that the human mind is capable of imagining are numerous, but many of those things never find their way into use, because of cognitive and functional constraints. Only a small set of "stable engineering solutions satisfying multiple design constraints" (Evans and Levinson 2009: 1) survive and thrive—because they are solutions to human problems, and because they are perceived as sufficiently relevant to warrant their adoption. There is such an alternative numeral system, and you've likely seen it before if you've ever looked inside electronics, although you don't necessarily know what it all means (figure 7.2).

Almost every electronic device uses resistors, like the one shown here to create resistance (measured in ohms) to the flow of electric current in the device, and they almost all have a set of colored bands to indicate their resistance, standardized internationally. The first two or three bands represent some numerical quantity, using different colors to indicate the numerals 0 through 9. So, in figure 7.2, the leftmost band is blue, representing 6, and the next is gray, for 8. The third band is the multiplier band—it uses the same color system, but represents powers of 10—so in this case, brown is 1, or $10^1 = 10$. The product of the first two bands and the multiplier, 68×10, gives the resistance, 680 ohms. The fourth band uses a different color scheme to represent tolerance, in this case gold meaning ±5%. It also serves a useful function of highlighting the end of the numeral—to quickly show the reader to start reading from the other end of the resistor.

The other advantage they have is that color bands are highly distinct and (except to the color-blind) visually salient at small scale—resistors are generally only a few millimeters thick and it is not always feasible to print highly readable numerals on or near them. The colors are useful for human eyes in the context of a nonlinear text medium (resistors can be oriented all sorts of ways within a circuit) where bright color visible at small scale can be readily processed. Cross-culturally, numerical notations almost never use color directly to represent differences in numerical value (Chrisomalis

Figure 7.2
680-ohm resistor using electronic color code (680 ohms 5% axial resistor by bomazi is licensed under CC BY-SA 2.0; source: Wikimedia Commons)

Table 7.2
Resistor numeral values

Color	Digit value	Multiplier value	Tolerance
Black	0	1	—
Brown	1	10	±1%
Red	2	100	±2%
Orange	3	1,000	—
Yellow	4	10,000	—
Green	5	100,000	±0.5%
Blue	6	1,000,000	±0.25%
Purple	7	10,000,000	± 0.1%
Gray	8	—	—
White	9	—	—
Gold	—	0.1	±5%
Silver	—	0.01	±10%

2010: 365). It is probable that color serves some semantic function in the encoding of the Inka *khipu* records (Hyland 2017), but not, apparently, to indicate differences in numerical value. But in most written traditions—whether incised in wood or stone, imprinted in clay, or written with ink on some flat surface—using color to denote semantic difference is rarely necessary. Without color, resistor numerals are structurally similar to someone saying "I make 56K a year"—56 being the digits, K being the multiplier for 1,000. They combine a ciphered-positional structure for the significant digits and then are multiplicative-additive for the power. But unlike "K," which is the only multiplier used in English in this way (as discussed in chapter 6), resistor numerals have a color band for each power, equal to the exponent of 10 being represented.

Resistor numerals are really most useful for representing round numbers—those with one or more trailing zeroes—in a very concise way. In this way they are similar to scientific notation, which is principally used to express large or small round numbers (6.02×10^{23}) but is fairly useless for numbers without lots of zeroes. To write 478,239 on a resistor would require six bars of different colors—4, 7, 8, 2, 3, and 9, followed by the multiplier bar for "×1." But no one makes a 478,239-ohm resistor—that kind of precision isn't needed most of the time. You might have a 470,000-ohm resistor, though, in which case you just need the 4 and 7, along with the multiplier for 10,000. Thus, resistors rarely need more than two or at most three "digit" bands, plus a multiplier band, to indicate the resistance—even for really high resistances up into the millions of ohms, which are actually quite common in everyday use. In many cases conciseness in numeration is not an issue, as I have shown repeatedly, but on a tiny resistor, being able to convey the resistance in just a couple of visually salient colored bands that are hard to misread is very valuable indeed.

Again, I don't suppose that everyone in a century or two is going to use color-coded bands for digits and a multiplier for representing numbers generally. Nonetheless, you have, in your home, numerous instances (possibly hundreds) of a radically different numeration system from the Western numerals, one that was invented in the mid twentieth century and is still used decades later in electronic devices throughout the world, albeit hidden from ready view. People—real people who design, install, and repair electronics—find them more useful than Western numerals for this purpose. We don't need to be transhuman to use them—they were designed by humans to be read by humans.

Not only are there imaginable alternatives to Western numerals, but some of them are actually adopted and in use. We should bear in mind that many new innovations will fall under the pressure of frequency-dependent biases, as we saw in the case of the Cherokee numerals. Even so, the fact that invention in the domain of number continues apace—and perhaps has even increased in pace, although that would be harder to show—demonstrates the continued vitality of human numerical inventiveness. We should thus feel liberated to ask under what conditions and in what contexts a seemingly ubiquitous "killer" system like the Western numerals can have its triumph upended. Knowing how it came to be universal is one step. These conditions are, I have shown, likely to be a combination of cognitive-structural and social ones, linked through ongoing processes of reckoning—evaluating, thinking, judging, deciding. The next step is to ask how this system came to be *believed* to be irrevocably triumphant. We can hardly foresee the details, but we should check our arrogant assumption that the narrative has ended forever. We are not at the end of history of numbers, and we never will be.

What are the limits on human variation?

In his startling essay *Fragments of an Anarchist Anthropology*, David Graeber sets out a forthright agenda preparing anthropology to aid in producing forms of sociopolitical anarchism (Graeber 2004). No armchair theorist, Graeber's goals are explicitly political: to demonstrate the feasibility of alternative political and economic solutions to those currently prevalent, and to actualize processes that will bring them about. Graeber and I share a common concern that anthropology is the only discipline well suited to examine the totality of human behavior and social organization, the only discipline whose explicit comparativism permits an honest investigation of the nature of inequality, violence, power, hierarchy, knowledge, and the state, and the interactions among them. We share a conviction that the sociologists, philosophers, and literary theorists who have dealt with these questions lately have done so poorly, and from a Western-centered viewpoint. It is very easy for those living in modern Western societies where states are large, powerful, and oppressive to imagine that there are no alternatives. Graeber's vision of an anthropology that does not yet exist is a discipline that seeks to present such alternatives, while recognizing that not everything is possible—and that it is thus imperative to circumscribe the possible within the imaginable.

Graeber also moves explicitly to include historical and archaeological time scales in his work—for instance, in his account of debt over the past 5,000 years (Graeber 2011). This is not to deny the value of the study of living people in all their complexity. Ethnography is one of the methods by which I and some other anthropologists produce contingent knowledge about present societies through sustained participant observation and other systematic interaction with living people. However, it is quite dangerous to equate the range of variability in human behavior today with the range that has existed in the past, or the range that might exist in the future. We run the risk of failing to observe social formations that once existed but no longer do, anywhere, and thus of constraining our ability to think about alternative solutions. In analyzing how societies resist state institutions, we must go beyond those ways used by people living under and resisting a particular form of domination, that which is particular to the past several centuries. These are five-hundred-year solutions to ten-thousand-year problems.

We do not know, nor is there any immediate prospect of knowing within any contemporary body of anthropological theory, what the constraints are on configurations of social inequality in human societies. Most of the literature I have discussed, including the numerical evidence, has identified constraints in cognitive and linguistic domains, not social ones, but that is largely because social constraints have not been conceptualized as such, rather than because they do not exist. Assuming that there are none evokes without warrant the specter of dystopian futures as much as it allows the prospect of utopian ones. It is imperative, if anthropology is to make any contribution to the social sciences in the twenty-first century, to answer the question, "To what degree, and by what processes, is variability in the degree and intensity of social inequality created?" Such an enterprise requires not only that we understand the range of variability in contemporary inequality but that we remind ourselves that this range is not fixed chronologically.

David Aberle's (1987) distinguished lecture "What Kind of Science Is Anthropology?" outlines a historical theory of anthropological constraints, rightly noting that "the historical constraints on a system are ever-changing, since some of the novelties of today that are incorporated into the system become the constraints of tomorrow" (Aberle 1987: 554). Recognizing that environmental and functional constraints (including cognitive ones) do play some role in constraining human behavior, Aberle insists that although anthropology cannot be a predictive science, it can and should be a probabilistic, explanatory historical science along with geology, historical linguistics, evolutionary biology, and cosmology. I agree with this fully. His insistence on the value of historical reconstructions using synchronic ethnological data is appropriate, but reconstructions that are not aware of the possibility that the past may be different from the present will almost surely be flawed. It is as if we were to argue for evolutionary taxonomy without paleontology even where the fossil record is abundant. Where historical or archaeological data are available, it is appropriate to use them to reconstruct the past in a more direct, less inferential manner. And so, while anthropology must be comparative, it must be comparative in its totality, across many time scales: ethnographic, historical, archaeological, evolutionary.

This is a call, then, for a macro-anthropology to parallel our current attention to micro-anthropology. Macrohistorical scholarship, for many historians, raises the specter of historians such as Arnold Toynbee (1934–1961) and Oswald Spengler (1926) who, copious though their work may be, lack the rigor that characterizes thorough historical scholarship. The challenge of such work is principally that its scope is too broad, seeking not only to unify all of world history but every subject of world history in grand narratives or epoch-spanning cycles. I do not claim, just because numeral systems are amenable to comparative analysis, that every subject or domain of experience must be as well, or that the history of numerals helps explain the history of all domains. I do insist, however, that anthropological theory ought to be grounded in the broadest range of data we can have available, and that that includes evidence from all historical periods.

Over the past century and a half, anthropology has been no stranger to asking big questions—theoretical ones that drive forward the discipline, even if unanswered, simply because they are asked. Yet anthropology over the past several decades has retreated from asking these big questions in favor of the local, the historically situated, and the contextual, overshadowing the need for anthropologists to develop and use theories of culture and behavior. This timidity is understandable as a reaction to racist and colonialist excesses, but it denies anthropology any reentry into relevance in understanding the human condition. As a result, anthropological contributions to the human sciences have been limited in a way that would have been unthinkable fifty years ago.

While we are living in the here and now, we are part of much larger-scale and longer-term processes: the "long now" that encompasses all the variability in behavior and knowledge of our species over the millennia of its existence. This phrase, coined by the musician and futurist Brian Eno, a founder of the Long Now Foundation, emphasizes the value of the macro scale to our understanding of the present (Brand 1999: 28). Anthropologists should recoil at the proposition that the local and the particular are all that we do well. Anthropology is the only discipline that claims to be able to study humanity at any time and in any place, in all its variability and sameness. We are gravely in need of a theory of power and inequality, one built by anthropologists with all the data that we are willing and able to gather.

This, for me, is the strongest rationale for the broad, historically inclusive, evolutionarily informed formulation of anthropology that has predominated in North America for the past century, and which ultimately has much deeper origins in Enlightenment and early evolutionary analyses of human behavior (Balée 2009). It is a call for holism; not, as Roy Ellen (2010) argues perceptively, for a vague meaningless undivided presumption of unity, but for a methodologically well-supported rejection of disunity among the subfields and topical specialties. I cannot see how "unwrapping the sacred bundle" is anything but detrimental to what anthropology has to offer the social sciences (Segal and Yanagisako 2005). To this formulation I would add historical anthropology, much-neglected yet crucial for integrating past and present. An anthropology that is ethnography and nothing more is unlikely to be relevant for explaining social configurations and remedying social problems beyond the local and contemporary.

Similarly, if anthropology is only a borrower rather than a lender of theory, a discipline whose social theories are borrowed rather than built from our data, then it will cease to have much relevance in the eyes both of other social scientists and of the general public. In this I share with the social anthropologist Matti Bunzl a concern that anthropology's lack of generalizing focus renders us irrelevant (Bunzl 2008). But the problem is not that disciplinary trends swing like a pendulum between generalizing and particularizing. Generalization has been out of fashion in anthropology for two generations now. The real problem is that we have reified this dichotomy and forgotten about diachronic cross-cultural comparison. Because most universalists and particularists presume that the ethnographic record will be a good basis for theorizing—for universalists because of the astonishing sameness of humanity, and for particularists because the past is no more and no less unique than the present—they neglect sources of data that are diachronically situated. An anthropology that aims to solve human problems cannot restrict itself to a tiny sliver of human variability.

Numerical notations are a particularly tractable domain of experience for this perspective. We have ample textual and archaeological evidence for their use over five thousand years of recorded history, and tantalizing evidence such as tallying going back tens of thousands of years earlier, into the Upper Paleolithic. Their materiality offers us a foothold into their contexts of use (Overmann 2016). They vary, but they do so in constrained ways, with the same five basic structures recurring multiple times independently, so they are neither so universal nor so variable as to be uninteresting. They have understandable histories—trajectories of development, use, and abandonment—that allow us some insight into more general cultural-evolutionary processes of long-term change. They have fruitful and persistent connections to language, a domain for which there is already a well-accepted framework for historical analysis (first philology, and now its descendant, historical linguistics). And they reflect a key interest in cognitive science—numerical cognition—where the study of links between language, notation, perception, and memory are over half a century old (Miller 1956). But I do not think, despite these advantages, that numerals are the only or even the principal subject amenable to this kind of analysis.

Broadly comparative and historical approaches to anthropological data are available, if we have the courage to try them.

In his last words published during his lifetime, the archaeologist Bruce Trigger said in an interview, "I look forward to the day when knowledge of human behavior has reached the point where archaeologists are not only able to understand social variation in the past but can help to construct credible models of societies that have never (yet) existed, in order to broaden and inform public discussion of future alternatives" (in Yellowhorn 2006: 324). Here I think Trigger was exactly right. I would expand his point to include all social scientists, not only archaeologists. We must not think that the present is the best guide to the future, just because yesterday is more distant from tomorrow than today is. Change counts, and a holistic, comparative, and historical anthropology is witness to diachronic and social processes at a multimillennial scale. An anthropology that aims to be relevant to contemporary life must, therefore, build on its foundations as a science of constraints, and to seek to become an anthropology of change.

Notes

Chapter 1 / I

1. I do not know what to make of the contradiction between "He had not written it down ..." and "Each word had a particular sign, a kind of mark ...". The resolution does not appear to be a translation issue (orig. "No lo había escrito ..." and "Cada palabra tenía un signo particular, una especie de marca; las últimas eran muy complicadas ...").

2. Interesting parallels for these sorts of specialized, materialized counting practices are found or have been used historically in many parts of the world, such as in Polynesia (Bender and Beller 2006) and among the Elem Pomo of California (Ahlers 2012).

3. But note that there are English words like *googol* (1 followed by 100 zeroes) that, if considered part of the numeral system (which Greenberg would not), do violate this generalization—there is no well-formed or accepted number word for "one less than a googol."

4. The use of subtractive notation, as in Roman MCMXCIX for 1,999, does not violate this principle because the powers (thousands—M, hundreds—CM, tens—XC, ones—IX) are still in order even though, within each power, the signs are inverted. If one could write ICILMXIXC for the Roman numeral 1,273, this would be an exception.

5. See, however, Andersen 2005 for a counterargument that sees numerals as historically more recent than other word categories.

6. Or, as Brooks and Wiley (1988: 103) call it, slightly more optimistically, "survival of the adequate."

7. The literature on the typology of writing systems is enormous and too peripheral to the present discussion to be covered here. See especially Trigger 1998 and Daniels and Bright 1996 for more information.

8. Unlike East Asian "abaci," which use beads on rods, Greco-Roman abaci were boards on which grooves or lines were inscribed, onto which pebbles or stones were

placed and manipulated. While the principle is generally the same, the ability to move stones from one column to another makes the use of the Greco-Roman technology quite distinct.

9. This perplexity is demonstrated by the European adoption of the Arabic word *sifr* "zero" not only as *zero*, but also as *cipher* "secret mode of writing."

Chapter 2 / II

1. This practice (numeral delimiter commas) only became customary in English texts in the eighteenth century; in traditions (such as in France) where a comma is used in place of the decimal point (e.g., 142,63 instead of 142.63), comma separation at the thousands and millions never became common practice. And in India, because languages like Hindi have special morphemes for *hundred thousand* and *ten million*, the comma structure is different: 1,00,000 instead of 100,000, for instance.

2. Incidentally, the forms of the half-powers of each of these numbers is taken from a literal halving of the sign—so, the right half of ↀ for 1,000 becomes D for 500. In fact, this principle is extended to L (the bottom half of C = 100) and V (the top half of X = 10). Although they look like letters, and clearly came to be understood as letters by later Romans, their visual etymology was remarkably iconic and nonalphabetic, drawing from a very different, and more ancient, graphic principle ultimately deriving from tallying practices (Keyser 1988).

Chapter 3 / III

1. In these exercises there is also confusion as to whether to present Roman numerals with the bottoms pointing toward the center of the clock face (as is traditional), toward the bottom of the clock (vertically), or even to switch between orientations depending on the number.

2. The exceptions are systems like the Cistercian semicryptographic numerals described in scholarship by David King (2001), in which the "positions" are not in a line but rotate around four corners of a fixed center point, and so are both positional and noninfinite.

3. There are also solutions such as scientific notation (10^{17}).

4. In general, the *suan pan* historically had five round beads in the bottom half of the device and two in the top, while the *soroban* had four diamond-shaped beads in the bottom and one in the top. But the differences in their use are minor, and this is not a hard and fast rule.

5. We might even see this as embodied cognition, as even able arithmeticians may calculate with "virtual paper" with a finger extended in front of the face, as if writing numbers in the air.

6. The relationship between metalanguage and the narrower concept of language ideology is significantly too complex to cover here. See, especially, Woolard and Schieffelin 1994 for more discussion.

7. I owe much of this analysis to productive discussions with John Dagenais about Llull's mathematics and the translations of the specific terms he was using.

8. In referring to the abacus, Llull was probably not talking about a computation board or device, because in the thirteenth century phrases like *figures de l'abaco* "abacus numerals" had, ironically enough, come to be applied to Western numerals, specifically under the influence of Fibonacci's *Liber abaci.*

9. Note that for 4 Aubrey uses iiii, not iv.

Chapter 6 / VI

1. As mentioned in chapter 2, Indian users of Western numerical notation usually divide long chunks of numerals differently than Western writers do—where I would write 100,000, a Hindi speaker might write 1,00,000.

2. The cover image of this book, a watch designed by Jean-Antoine Lépine in 1788, is a similarly playful blended modality of two numerical notations (Roman and Western numerals). Thompson (2008: 100–101) suggests that this may have been to keep the watch face uncluttered, as each number takes at most two digits.

3. There is probably no better way 4 U 2 seem hopelessly old than 2 try 2 imit8 the inventive online linguistic practices of youth, which change extremely rapidly (Tagliamonte 2016).

4. Compare with modern Arabic ستة الاف *stt alaf.*

5. Here I'm conforming to the style guide—rather than just start the sentence "417," which wouldn't do, I am enjoined to rewrite it so as not to have to violate a rule.

6. In June 2020, the technology entrepreneur Elon Musk and his partner, the musician Grimes, had a baby boy. Originally the plan was to name the young lad X Æ A-12. It turns out, though, that this violates California law which mandates that numerals cannot be used in a legal name. Their solution was to name the child X AE A-XII instead—using Roman numerals, because they apparently count as letters rather than numbers. But when asked about her choice, Grimes indicated that the new name looked better anyway.

Chapter 7 / VII

1. This may have had a greater impact on you than you imagine. Some research shows that among individuals with *color-grapheme synesthesia*—people who perceive letters and numerals as having inherent colors—the specific associations they had

were strongly correlated with the most popular colored letter and numeral toys widely available in their infancy (Witthoft and Winawer 2006, 2013).

2. Most of these are in fact quinary-vigesimal (mixed base-5 and base-20), or decimal-vigesimal (mixed base-10 and base-20). Only a few languages have totally distinct words for 1 through 20 without any subdivisions or structures.

3. Hammarström (2010: 32) also reports that historically, though no longer, Ntomba (a Bantu language of the Democratic Republic of the Congo) had a base-60 structure.

4. Despite the time and distance of the purported connection, this was a serious debate for a time in the 1970s, occupying considerable attention, as charges of racism flew in both directions among different parties to the dispute (Bowers and Lepi 1975; Pospisil and Price 1976; Bowers 1977).

5. A notable exception is Ursula K. Le Guin's first published science fiction story, "The Masters" (first published in 1961), in which a six-fingered future (or past) human species uses base-12 cumulative-additive numerals, much like the Roman numerals, but where the invention of a zero and ciphered-positional notation brings chaos (Le Guin 1975). I do not think it is a coincidence, in mentioning this rather obscure early work, that Le Guin was the daughter of A. L. Kroeber and was heavily influenced by his anthropology.

6. Just before this volume went to press, however, a new preprint has suggested that the Maya numerical notation was originally a bijective positional system where the sign for 0 was actually a sign for 20 (Rojo-Garibaldi et al. 2020). I'm doubtful (no archaeologists or linguists were involved in the study), but it is not inherently implausible.

Bibliography

Aberle, David F. 1987. Distinguished Lecture: What Kind of Science Is Anthropology? *American Anthropologist* 89 (3): 551–566.

Adler, Jerry. 1954. So You Think You Can Count! *Mathematics Magazine* 28 (2): 83–86.

Ahearn, Laura M. 2001. Language and Agency. *Annual Review of Anthropology* 30 (1): 109–137.

Ahlers, Jocelyn C. 2012. Two Eights Make Sixteen Beads: Historical and Contemporary Ethnography in Language Revitalization. *International Journal of American Linguistics* 78 (4): 533–555.

Allen, James P. 2000. *Middle Egyptian: An Introduction to the Language and Culture of Hieroglyphs*. Cambridge: Cambridge University Press.

Andersen, Henning. 2005. The Plasticity of Universal Grammar. In *Convergence: Interdisciplinary Communications 2004/2005*, edited by W. Østreng, 21–26. Oslo: Center for Advanced Studies at the Norwegian Academy of Sciences and Letters.

Anderson, W. French. 1956. Arithmetical Computations in Roman Numerals. *Classical Philology* 51 (3): 145–150.

Andrews, Frank Emerson. 1944. *New Numbers: How Acceptance of a Duodecimal (12) Base Would Simplify Arithmetic*. New York: Essential Books.

Anonymous. 1910. "Abacus Beats an Adding Machine." *Westminster (MD) Democratic Advocate*, July 22, 1910, 5.

Anonymous. 1941. *Industrial Canada* 42 (1–6): 154.

Anonymous. 1949. Report on U.S. Dry Cargo Overseas Trade. *The Log* 44 (8): 80–86.

Anonymous. 1953. Official Proceedings of the National Guard Association of the United States. San Diego, CA.

Arbuthnot, John. 1701. *An Essay on the Usefulness of Mathematical Learning: In a Letter from a Gentleman in the City to His Friend in Oxford*. Oxford: A. Peisley.

Arthur, W. Brian. 1990. Positive Feedbacks in the Economy. *Scientific American* 262 (2): 92–99.

Artioli, Gilberto, Viviana Nociti, and Ivana Angelini. 2011. Gambling with Etruscan Dice: A Tale of Numbers and Letters. *Archaeometry* 53 (5): 1031–1043.

Asch, Solomon E. 1955. Opinions and Social Pressure. *Scientific American* 193 (5): 31–35.

Aubrey, John. 1881. *Remaines of Gentilisme and Judaisme*. London: W. Satchell.

Babu, Senthil. 2007. Memory and Mathematics in the Tamil *Tinnai* Schools of South India in the Eighteenth and Nineteenth Centuries. *International Journal for the History of Mathematics Education* 2 (1): 15–32.

Bailey, Melissa. 2013. Roman Money and Numerical Practice. *Revue belge de philologie et d'histoire* 91 (1): 153–186.

Baines, John. 1989. Communication and Display: The Integration of Early Egyptian Art and Writing. *Antiquity* 63 (240): 471–482.

Baines, John. 2007. *Visual and Written Culture in Ancient Egypt*. Oxford: Oxford University Press.

Baines, John, John Bennet, and Stephen Houston. 2011. *The Disappearance of Writing Systems: Perspectives on Literacy and Communication*. London: Equinox.

Balée, William. 2009. The Four-Field Model of Anthropology in the United States. *Amazônica, Revista de Antropologia* 1 (1). https://doi.org/http://dx.doi.org/10.18542/amazonica.v1i1.136.

Barany, Michael J. 2014. Savage Numbers and the Evolution of Civilization in Victorian Prehistory. *British Journal for the History of Science* 47 (2): 239–255.

Barnett, Homer G. 1953. *Innovation: The Basis of Cultural Change*. New York: McGraw-Hill.

Bartley, William Clark. 2002. Counting on Tradition: Iñupiaq Numbers in the School Setting. In *Perspectives on Indigenous People of North America*, edited by J. E. Hankes and G. R. Fast, 225–236. Reston, VA: National Council of Teachers of Mathematics.

Bauman, Richard. 2008. The Philology of the Vernacular. *Journal of Folklore Research* 45 (1): 29–36.

Bazzanella, Marta, Giovanni Kezich, and Luca Pisoni. 2014. "Adio pastori!" Ethics and Aesthetics of an Alphabetized Pastoral Subculture: The Case of Fiemme in the Eastern Alps (1680–1940). *Boletín del Museo Chileno de Arte Precolombino* 19 (1): 23–35.

Bender, Andrea, and Sieghard Beller. 2006. "Fanciful" or Genuine? Bases and High Numerals in Polynesian Number Systems. *Journal of the Polynesian Society* 115 (1): 7–46.

Bender, Andrea, Dirk Schlimm, and Sieghard Beller. 2015. The Cognitive Advantages of Counting Specifically: A Representational Analysis of Verbal Numeration Systems in Oceanic Languages. *Topics in Cognitive Science* 7 (4): 552–569.

Bender, Margaret Clelland. 2002. *Signs of Cherokee Culture: Sequoyah's Syllabary in Eastern Cherokee Life*. Chapel Hill: University of North Carolina Press.

Benefiel, Rebecca R. 2010. Dialogues of Ancient Graffiti in the House of Maius Castricius in Pompeii. *American Journal of Archaeology* 114 (1): 59–101.

Bentley, R. Alexander. 2006. Academic Copying, Archaeology and the English Language. *Antiquity* 80 (307): 196–201.

Bi Bo and Nicholas Sims-Williams. 2010. Sogdian Documents from Khotan, I: Four Economic Documents. *Journal of the American Oriental Society* 130 (4): 497–508.

Biella, Joan Copeland. 1982. *Dictionary of Old South Arabic, Sabaean Dialect*. Chico, CA: Scholars Press.

Biggs, Robert D., and Matthew W. Stolper. 1983. A Babylonian Omen Text from Susiana. *Revue d'assyriologie et d'archéologie orientale* 77 (2): 155–162.

Blades, William. 1882. *The Biography and Typography of William Caxton: England's First Printer*. Strasbourg: Trübner.

Blixen, Karen. 1937. *Out of Africa*. London: Penguin Books.

Boas, Franz. 1896. The Limitations of the Comparative Method of Anthropology. *Science* 4 (103): 901–908.

Borges, Jorge Luis. 1964. *Labyrinths: Selected Stories and Other Writings*. New York: New Directions.

Bostrom, Nick. 2014. *Superintelligence: Paths, Dangers, Strategies*. Oxford: Oxford University Press.

Bowers, Nancy. 1977. Kapauku Numeration: Reckoning, Racism, Scholarship, and Melanesian Counting Systems. *Journal of the Polynesian Society* 86 (1): 105–116.

Bowers, Nancy, and Pundia Lepi. 1975. Kaugel Valley Systems of Reckoning. *Journal of the Polynesian Society* 84 (3): 309–324.

Boyd, Robert, and Peter J. Richerson. 1985. *Culture and the Evolutionary Process*. Chicago: University of Chicago Press.

Boyer, Carl B. 1944. Fundamental Steps in the Development of Numeration. *Isis* 35 (2): 153–168.

Brand, Stewart. 1999. *The Clock of the Long Now: Time and Responsibility*. New York: Basic Books.

Brooks, Daniel R., and Edward O. Wiley. 1988. *Evolution as Entropy: Toward a Unified Theory of Biology*. Chicago: University of Chicago Press.

Brost, Gerard Robert. 1989. Dyhexal Numbers: How They Facilitate Arithmetic. *International Journal of Mathematical Education in Science and Technology* 20 (2): 249–253.

Brown, Cecil H. 1976. General Principles of Human Anatomical Partonomy and Speculations on the Growth of Partonomic Nomenclature. *American Ethnologist* 3 (3): 400–424.

Brown, Cecil H., and Stanley R. Witkowski. 1981. Figurative Language in a Universalist Perspective. *American Ethnologist* 8 (3): 596–615.

Brown, Donald E. 1991. *Human Universals*. Philadelphia: Temple University Press.

Budge, E. A. Wallis. 1911. *Osiris and the Egyptian Resurrection*. London: P. L. Warner; New York: G. P. Putnam's Sons.

Bunzl, Matti. 2008. The Quest for Anthropological Relevance: Borgesian Maps and Epistemological Pitfalls. *American Anthropologist* 110 (1): 53–60.

Burnett, Charles. 1997. *The Introduction of Arabic Learning into England*. The Panizzi Lectures, 1996. London: British Library.

Burnett, Charles. 2000. Latin Alphanumerical Notation, and Annotation in Italian, in the Twelfth Century: MS London, British Library, Harley 5402. In *Sic itur ad astra: Studien zur Geschichte der Mathematik und Naturwissenschaften*, edited by Menso Folkerts and Richard Lorch, 76–90. Wiesbaden: Harrassowitz.

Burnett, Charles. 2002. Learning Indian Arithmetic in the Early Thirteenth Century. *Boletín de la Asociación Matemática Venezolana* 9 (1): 15–26.

Burnett, Charles. 2006. The Semantics of Indian Numerals in Arabic, Greek and Latin. *Journal of Indian Philosophy* 34 (1–2): 15–30.

Burnett, Charles, and William F. Ryan. 1988. Abacus (Western). In *Instruments of Science: An Historical Encyclopedia*, edited by Robert Bud, 5–7. London: Science Museum.

Byock, Jesse L. 1990. *Medieval Iceland: Society, Sagas, and Power*. Berkeley: University of California Press.

Carey, Susan. 2009. *The Origin of Concepts*. Oxford: Oxford University Press.

Carroll, Beau Duke, Alan Cressler, Tom Belt, Julie Reed, and Jan F. Simek. 2019. Talking Stones: Cherokee Syllabary in Manitou Cave, Alabama. *Antiquity* 93 (368): 519–536.

Carvalho, Joaquim Barradas de. 1957. Sur l'introduction et la diffusion des chiffres arabes au Portugal. *Bulletin des études portugais* 20: 110–151.

Cattell, J. McKeen. 1914. Science, Education and Democracy. *Science* 39 (996): 154–164.

Chadwick, John. 1990. *The Decipherment of Linear B*. New York: Cambridge University Press.

Chrisomalis, Stephen. 2003. The Egyptian Origin of the Greek Alphabetic Numerals. *Antiquity* 77 (297): 485–496.

Chrisomalis, Stephen. 2007. The Perils of Pseudo-Orwellianism. *Antiquity* 81 (311): 204–207.

Chrisomalis, Stephen. 2009. The Origins and Coevolution of Literacy and Numeracy. In *The Cambridge Handbook of Literacy*, edited by David R. Olson and Nancy Torrance, 59–74. Cambridge: Cambridge University Press.

Chrisomalis, Stephen. 2010. *Numerical Notation: A Comparative History*. New York: Cambridge University Press.

Chrisomalis, Stephen. 2013. Constraint, Cognition, and Written Numeration. *Pragmatics and Cognition* 21 (3): 552–572.

Chrisomalis, Stephen. 2016. Umpteen Reflections on Indefinite Hyperbolic Numerals. *American Speech* 91 (1): 3–33.

Chrisomalis, Stephen. 2017. Re-evaluating Merit: Multiple Overlapping Factors Explain the Evolution of Numerical Notations. *Writing Systems Research* 9 (1): 1–13.

Clark, Andy. 2008. *Supersizing the Mind: Embodiment, Action, and Cognitive Extension*. Oxford: Oxford University Press.

Cohen, Patricia Cline. 1982. *A Calculating People: The Spread of Numeracy in Early America*. Chicago: University of Chicago Press.

Comrie, Bernard, and David Gil. 2005. *The World Atlas of Language Structures*. Vol. 1. Oxford: Oxford University Press.

Coupland, Nikolas. 2011. How Frequent Are Numbers? *Language and Communication* 31 (1): 27–37.

Crawfurd, John. 1863. On the Numerals as Evidence of the Progress of Civilization. *Transactions of the Ethnological Society of London* 2: 84–111.

Crisostomo, C. Jay. 2015. Language, Writing, and Ideologies in Contact: Sumerian and Akkadian in the Early Second Millennium BCE. In *Semitic Languages in Contact*, edited by Aaron Butts, 158–180. Leiden: Brill.

Crossley, John. 2013. Old-Fashioned versus Newfangled: Reading and Writing Numbers, 1200–1500. *Studies in Medieval and Renaissance History* 10: 79–109.

Crystal, David. 2001. *Language Play*. Chicago: University of Chicago Press.

Cushman, Ellen. 2012. *The Cherokee Syllabary: Writing the People's Perseverance*. Norman: University of Oklahoma Press.

Dagenais, John, trans. 2019. *Doctrina pueril: A Primer for the Medieval World*. Woodbridge, UK: Boydell and Brewer.

Damerow, Peter. 1996. *Abstraction and Representation: Essays on the Cultural Evolution of Thinking*. Dordrecht: Kluwer Academic Publishers.

Daniels, Peter T., and William Bright. 1996. *The World's Writing Systems*. New York: Oxford University Press.

Darwin, Charles. 1859. *On the Origin of Species by Means of Natural Selection, or the Preservation of Favoured Races in the Struggle for Life*. London: John Murray.

Dawdy, Shannon Lee. 2010. Clockpunk Anthropology and the Ruins of Modernity. *Current Anthropology* 51 (6): 761–793.

De Cruz, Helen. 2006. Why Are Some Numerical Concepts More Successful Than Others? An Evolutionary Perspective on the History of Number Concepts. *Evolution and Human Behavior* 27 (4): 306–323.

De Cruz, Helen, Hansjörg Neth, and Dirk Schlimm. 2010. The Cognitive Basis of Arithmetic. In *PhiMSAMP. Philosophy of Mathematics: Sociological Aspects and Mathematical Practice*, edited by B. Löwe and T. Müller, 59–106. London: College Publications.

DeFrancis, John. 1984. *The Chinese Language: Fact and Fantasy*. Honolulu: University of Hawaii Press.

Dehaene, Stanislas. 1997. *The Number Sense: How the Mind Creates Mathematics*. Oxford: Oxford University Press.

d'Errico, Francesco. 1998. Palaeolithic Origins of Artificial Memory Systems: An Evolutionary Perspective. In *Cognition and Material Culture: The Archaeology of Symbolic Storage*, edited by Colin Renfrew and Chris Scarre, 19–50. Cambridge: McDonald Institute for Archaeological Research.

d'Errico, Francesco, and Carmen Cacho. 1994. Notation versus Decoration in the Upper Palaeolithic: A Case-Study from Tossal de la Roca, Alicante, Spain. *Journal of Archaeological Science* 21 (2): 185–200.

Derzhanski, Ivan A. 2004. Codex Seraphinianus: Some Observations. http://www.math.bas.bg/~iad/serafin.html.

Dethlefsen, Edwin, and James Deetz. 1966. Death's Heads, Cherubs, and Willow Trees: Experimental Archaeology in Colonial Cemeteries. *American Antiquity* 31 (4): 502–510.

Detlefsen, Michael, Douglas K. Erlandson, J. Clark Heston, and Charles M. Young. 1976. Computation with Roman Numerals. *Archive for History of Exact Sciences* 15 (2): 141–148.

Devereux, Edward James. 1999. *Bibliography of John Rastell*. Montreal: McGill-Queen's University Press.

Dobres, Marcia-Anne, and John E. Robb, eds. 2000. *Agency in Archaeology*. London: Routledge.

Dryer, Matthew S. 1997. Why Statistical Universals Are Better than Absolute Universals. *Proceedings of the 33rd Annual Meeting of the Chicago Linguistic Society* 33: 123–145.

Dryer, Matthew S. 2003. Significant and Non-significant Implicational Universals. *Linguistic Typology* 7 (1): 108–128.

Durham, John W. 1992. The Introduction of "Arabic" Numerals in European Accounting. *Accounting Historians Journal* 19 (2): 25–55.

Durrenberger, E. Paul. 1992. *The Dynamics of Medieval Iceland: Political Economy and Literature*. Iowa City: University of Iowa Press.

Edzard, Dietz Otto. 1981. *Verwaltungstexte verschiedenen Inhalts aus dem Archiv L. 2769: ARET II*. Rome: Missione archeologica italiana in Siria.

Eells, Walter Crosby. 1913. Number Systems of the North American Indians. *American Mathematical Monthly* 20 (10): 293–299.

Efferson, Charles, Rafael Lalive, Peter J. Richerson, Richard McElreath, and Mark Lubell. 2008. Conformists and Mavericks: The Empirics of Frequency-Dependent Cultural Transmission. *Evolution and Human Behavior* 29 (1): 56–64.

Eisenstein, Elizabeth L. 1979. *The Printing Press as an Agent of Change: Communications and Cultural Transformations in Early-Modern Europe*. Cambridge: Cambridge University Press.

Elkins, James. 1996. On the Impossibility of Close Reading: The Case of Alexander Marshack. *Current Anthropology* 37 (2): 185–226.

Ellen, Roy. 2010. Theories in Anthropology and "Anthropological Theory." *Journal of the Royal Anthropological Institute* 16 (2): 387–404.

Emigh, Rebecca Jean. 2002. Numeracy or Enumeration? The Uses of Numbers by States and Societies. *Social Science History* 26 (4): 653–698.

Englehardt, Joshua, ed. 2012. *Agency in Ancient Writing*. Boulder: University Press of Colorado.

Evans, Arthur J. 1909. *Scripta Minoa*. Vol. 1. Oxford: Clarendon Press.

Evans, Nicholas, and Stephen C. Levinson. 2009. The Myth of Language Universals: Language Diversity and Its Importance for Cognitive Science. *Behavioral and Brain Sciences* 32 (5): 429–448.

Everett, Caleb. 2017. *Numbers and the Making of Us: Counting and the Course of Human Cultures*. Cambridge, MA: Harvard University Press.

Everett, Daniel L. 2005. Cultural Constraints on Grammar and Cognition in Pirahã: Another Look at the Design Features of Human Language. *Current Anthropology* 46 (4): 621–646.

Fales, Frederick Mario. 1995. Assyro-Aramaica: The Assyrian Lion-Weights. In *Immigration and Emigration within the Ancient Near East: Festschrift E. Lipinski*, edited by Karel van Lerberghe and Antoon Schoors, 33–55. Leuven: Peeters.

Ferrero, Guillaume. 1894. L'inertie mentale et la loi du moindre effort. *Revue philosophique de la France et de l'Étranger* 37: 169–182.

Ford, John C. 2018. Two or III Feet Apart: Oral Recitation, Roman Numerals, and Metrical Regularity in *Capystranus*. *Neophilologus* 102 (4): 573–593.

Forslund, Robert R. 1995. A Logical Alternative to the Existing Positional Number System. *Southwest Journal of Pure and Applied Mathematics* 1: 27–29.

Foster, James E. 1947. A Number System without a Zero-Symbol. *Mathematics Magazine* 21 (1): 39–41.

Foster, James E. 1953. Don't Call It Science. *Mathematics Magazine* 26 (4): 209–214.

Foster, James E. 1956. *Mathematics Need Not Be Practical*. Evanston: Principia Press of Illinois.

Frank, Michael C., Daniel L. Everett, Evelina Fedorenko, and Edward Gibson. 2008. Number as a Cognitive Technology: Evidence from Pirahã Language and Cognition. *Cognition* 108 (3): 819–824.

French, Christopher C., and Anne Richards. 1993. Clock This! An Everyday Example of a Schema-Driven Error in Memory. *British Journal of Psychology* 84 (2): 249–253.

Freund, John C. 1918. The Labor Supply. *The Music Trades*, December 14.

Friberg, Jöran. 2007. *A Remarkable Collection of Babylonian Mathematical Texts: Manuscripts in the Schøyen Collection: Cuneiform Texts I*. New York: Springer Science and Business Media.

Fukuyama, Francis. 1992. *The End of History and the Last Man*. New York: Free Press.

Gardner, Martin. 1956. *Mathematics, Magic and Mystery*. New York: Dover.

Gibson, Craig A., and Francis Newton. 1995. Pandulf of Capua's *De calculatione*: An Illustrated Abacus Treatise and Some Evidence for the Hindu-Arabic Numerals in Eleventh-Century South Italy. *Mediaeval Studies* 57: 293–335.

Gibson, James J. 1979. *The Ecological Approach to Visual Perception*. Boston: Houghton Mifflin.

Giddens, Anthony. 1979. *Central Problems in Social Theory: Action, Structure and Contradiction in Social Analysis*. Berkeley: University of California Press.

Goldenweiser, Alexander. 1913. The Principle of Limited Possibilities in the Development of Culture. *Journal of American Folklore* 26 (101): 259–290.

Goldenweiser, Alexander. 1942. *Anthropology: An Introduction to Primitive Culture*. New York: F. S. Crofts.

Gonzalez, Esther G., and Paul A. Kolers. 1982. Mental Manipulation of Arithmetic Symbols. *Journal of Experimental Psychology: Learning, Memory, and Cognition* 8 (4): 308–319.

Goody, Jack. 1977. *The Domestication of the Savage Mind*. Cambridge: Cambridge University Press.

Goody, Jack. 1982. *Cooking, Cuisine and Class: A Study in Comparative Sociology*. Cambridge: Cambridge University Press.

Goody, Jack. 1983. *The Development of the Family and Marriage in Europe*. Cambridge: Cambridge University Press.

Goody, Jack. 1986. *The Logic of Writing and the Organization of Society*. Cambridge: Cambridge University Press.

Goody, Jack, and Ian Watt. 1963. The Consequences of Literacy. *Comparative Studies in Society and History* 5 (3): 304–345.

Gordon, Arthur Ernest. 1983. *Illustrated Introduction to Latin Epigraphy*. Berkeley: University of California Press.

Gould, Stephen Jay, and Elisabeth S. Vrba. 1982. Exaptation—A Missing Term in the Science of Form. *Paleobiology* 8 (1): 4–15.

Graeber, David. 2004. *Fragments of an Anarchist Anthropology*. Chicago: Prickly Paradigm Press.

Graeber, David. 2011. *Debt: The First 5,000 Years*. New York: Melville House.

Graeber, David, and Marshall Sahlins. 2017. *On Kings*. Chicago: Hau Books.

Graham, Ian. 1997. *Corpus of Maya Hieroglyphic Inscriptions*, vol. 8, part 1: *Coba*. Cambridge, MA: Peabody Museum of Archaeology and Ethnology, Harvard University.

Gray, J. Patrick, and André Costopoulos. 2006. On Artificial Trends in Comparative Studies Using Standard Cross-Cultural Sample Data: Possibility and Probability. *Current Anthropology* 47 (1): 149–151.

Green, Margaret. 1949. Bacteriology of Shrimp, II: Quantitative Studies on Freshly Caught and Iced Shrimp. *Journal of Food Science* 14 (5): 372–383.

Greenberg, Joseph H. 1966. Synchronic and Diachronic Universals in Phonology. *Language* 42 (2): 508–517.

Greenberg, Joseph H. 1978. Generalizations about Numeral Systems. In *Universals of Human Language*, edited by Joseph H. Greenberg, 249–295. Stanford: Stanford University Press.

Grice, H. Paul. 1991. *Studies in the Way of Words*. Cambridge, MA: Harvard University Press.

Gronemeyer, Sven. 2004. A Preliminary Ruling Sequence of Cobá, Quintana Roo. *Wayeb Notes* 14. http://hdl.handle.net/20.500.11811/1102.

Hagerstrand, Torsten. 1967. *Innovation Diffusion as a Spatial Process*. Chicago: University of Chicago Press.

Hallpike, Christopher R. 1986. *The Principles of Social Evolution*. Oxford: Clarendon Press.

Hammarström, Harald. 2010. Rarities in Numeral Systems. In *Rethinking Universals: How Rarities Affect Linguistic Theory*, edited by Jan Wohlgemuth and Michael Cysouw, 11–53. Berlin: De Gruyter.

Hardy, Godfrey H. 1940. *A Mathematician's Apology*. London: Cambridge University Press.

Harris, Rollin A. 1905. Numerals for Simplifying Addition. *American Mathematical Monthly* 12 (3): 64–67.

Hastrup, Kirsten. 1985. *Culture and History in Medieval Iceland: An Anthropological Analysis of Structure and Change*. Oxford: Clarendon Press.

Hatano, Giyoo, and Keiko Osawa. 1983. Digit Memory of Grand Experts in Abacus-Derived Mental Calculation. *Cognition* 15 (1–3): 95–110.

Henrich, Joseph, Steven J. Heine, and Ara Norenzayan. 2010. The Weirdest People in the World? *Behavioral and Brain Sciences* 33 (2–3): 61–83.

Hewes, Gordon W. 1981. Prospects for More Productive Comparative Civilizational Studies. *Behavior Science Research* 16 (1–2): 167–185.

Hewes, Gordon W. 1995. The Daily Life Component in Civilization Analysis. *Comparative Civilizations Review* 33 (5): 79–96.

Hodgen, Margaret Trabue. 1945. Glass and Paper: An Historical Study of Acculturation. *Southwestern Journal of Anthropology* 1 (4): 466–497.

Hodgen, Margaret Trabue. 1952. *Change and History*. New York: Wenner-Gren Foundation for Anthropological Research.

Hodgen, Margaret Trabue. 1974. *Anthropology, History, and Cultural Change*. Tucson: University of Arizona Press.

Hoffecker, John F. 2007. Representation and Recursion in the Archaeological Record. *Journal of Archaeological Method and Theory* 14 (4): 359–387.

Hoffner, Harry A. 2007. On Higher Numbers in Hittite. *Studi micenei ed egeo-anatolici* 49 (1): 377–385.

Hofstra, Bas, Rense Corten, and Frank Van Tubergen. 2016. Who was First on Facebook? Determinants of Early Adoption among Adolescents. *New Media and Society* 18 (10): 2340–2358.

Holmes, Ruth Bradley, and Betty Sharp Smith. 1977. *Beginning Cherokee*. Norman: University of Oklahoma Press.

Houston, Stephen D. 1984. An Example of Homophony in Maya Script. *American Antiquity* 49 (4): 790–805.

Houston, Stephen D. 2004. The Archaeology of Communication Technologies. *Annual Review of Anthropology* 33: 223–250.

Houston, Stephen, and Andréas Stauder. 2020. What Is a Hieroglyph? *L'Homme* 233 (1): 9–44.

Hull, Glynda A., and Mark Evan Nelson. 2005. Locating the Semiotic Power of Multimodality. *Written Communication* 22 (2): 224–261.

Hutchins, Edwin. 1995. *Cognition in the Wild*. Cambridge, MA: MIT Press.

Hyland, Sabine. 2017. Writing with Twisted Cords: The Inscriptive Capacity of Andean Khipus. *Current Anthropology* 58 (3): 412–419.

Hymes, Dell H. 1963. Notes toward a History of Linguistic Anthropology. *Anthropological Linguistics* 5 (1): 59–103.

Ifrah, Georges. 1998. *The Universal History of Numbers*. London: Harvill Press.

Irvine, Judith T. 1989. When Talk Isn't Cheap: Language and Political Economy. *American Ethnologist* 16 (2): 248–267.

Jenkinson, Hilary. 1926. The Use of Arabic and Roman Numerals in English Archives. *Antiquaries Journal* 6 (3): 263–275.

Jervis, Ben. 2017. Assembling the Archaeology of the Global Middle Ages. *World Archaeology* 49 (5): 666–680.

Johnson, J. Cale. 2013. Indexical Iconicity in Sumerian *Belles Lettres*. *Language and Communication* 33 (1): 26–49.

Johnston, Alan W. 1979. *Trademarks on Greek Vases*. Warminster: Aris and Phillips.

Joos, Martin. 1936. Review: *The Psycho-biology of Language*. *Language* 12 (3): 196–210.

Kaufman, Edna L., Miles W. Lord, Thomas Whelan Reese, and John Volkmann. 1949. The Discrimination of Visual Number. *American Journal of Psychology* 62 (4): 498–525.

Kazem-Zadeh, Hossein. 1915. *Les chiffres siyâk et la comptabilité persane*. Paris: Ernest Leroux.

Keene, Bryan C. 2019. *Toward a Global Middle Ages: Encountering the World through Illuminated Manuscripts*. Los Angeles: Getty Publications.

Kennedy, James G. 1981. Arithmetic with Roman Numerals. *American Mathematical Monthly* 88 (1): 29–32.

Keyser, Paul. 1988. The Origin of the Latin Numerals 1 to 1000. *American Journal of Archaeology* 92 (4): 529–546.

Keyser, Paul T. 2015. Compound Numbers and Numerals in Greek. *Syllecta Classica* 26 (1): 113–175.

Kharsekin, Alexei I. 1967. On the Interpretation of Etruscan Numerals. *Soviet Anthropology and Archeology* 5 (3–4): 39–50.

King, David A. 2001. *The Ciphers of the Monks: A Forgotten Number-Notation of the Middle Ages*. Stuttgart: Franz Steiner Verlag.

Knapp, Samuel L. 1829. *Lectures on American Literature: With Remarks on Some Passages of American History*. New York: Elam Bliss.

Kojima, Takashi. 1954. *The Japanese Abacus: Its Use and Theory*. Rutland, VT: Charles E. Tuttle.

Kondratieff, Eric. 2004. The Column and Coinage of C. Duilius: Innovations in Iconography in Large and Small Media in the Middle Republic. *Scripta Classica Israelica* 23: 1–39.

Krenkel, Werner. 1969. Das Rechnen mit römischen Ziffern. *Das Altertum* 15: 252–256.

Kroeber, Alfred Louis. 1919. On the Principle of Order in Civilization as Exemplified by Changes of Fashion. *American Anthropologist* 21 (3): 235–263.

Kroeber, Alfred Louis. 1923. *Anthropology*. New York: Harcourt, Brace.

Krueger, Hilmar C. 1977. The Arabic Numerals of a Genoese Notary, 1202–1226. *Bollettino linguistico per la storia e la cultura regionale (Genoa)* 27 (3–4): 51–64.

Kunitzsch, Paul. 2003. The Transmission of Hindu-Arabic Numerals Reconsidered. In *The Enterprise of Science in Islam: New Perspectives*, edited by Jan P. Hogendijk and Abdelhamid I. Sabra, 3–21. Cambridge, MA: MIT Press.

Kurzweil, Ray. 2005. *The Singularity Is Near*. London: Penguin.

Kwon, Sang Jib, Eunil Park, and Ki Joon Kim. 2014. What Drives Successful Social Networking Services? A Comparative Analysis of User Acceptance of Facebook and Twitter. *Social Science Journal* 51 (4): 534–544.

Lakoff, George. 1987. *Women, Fire, and Dangerous Things: What Categories Reveal about the Mind*. Chicago: University of Chicago Press.

Lakoff, George, and Rafael E. Núñez. 2000. *Where Mathematics Comes From: How the Embodied Mind Brings Mathematics into Being*. New York: Basic Books.

Landy, David, Noah Silbert, and Aleah Goldin. 2013. Estimating Large Numbers. *Cognitive Science* 37 (5): 775–799.

Laroche, Roland A. 1977. Valerius Antias and His Numerical Totals: A Reappraisal. *Historia: Zeitschrift für alte Geschichte* 26 (3): 358–368.

Le Guin, Ursula K. 1975. The Masters. In *The Wind's Twelve Quarters*, 37–54. New York: Harper.

Lemay, Richard. 1977. The Hispanic Origin of Our Present Numeral Forms. *Viator* 8: 435–477.

Lemay, Richard. 1982. Arabic Numerals. In *Dictionary of the Middle Ages*, edited by Joseph R. Strayer, 382–398. New York: Scribner.

Lemay, Richard. 2000. Nouveautés fugaces dans des textes mathématiques du XIIe siècle: un essai d'abjad latin avorte. In *Sic itur ad astra: Studien zur Geschichte der Mathematik und Naturwissenschaften*, edited by Menso Folkerts and Richard Lorch, 376–389. Wiesbaden: Harrassowitz.

Levinson, Stephen C., and Nicholas Evans. 2010. Time for a Sea-Change in Linguistics: Response to Comments on "The Myth of Language Universals." *Lingua* 120 (12): 2733–2758.

Libbrecht, Ulrich. 1973. *Chinese Mathematics in the Thirteenth Century: The Shu-shu chiu-chang of Ch'in Chiu-shao*. Cambridge, MA: MIT Press.

Lieberman, Erez, Jean-Baptiste Michel, Joe Jackson, Tina Tang, and Martin A. Nowak. 2007. Quantifying the Evolutionary Dynamics of Language. *Nature* 449 (7163): 713–716.

Lloyd, Geoffrey E. R. 2007. *Cognitive Variations: Reflections on the Unity and Diversity of the Human Mind*. Oxford: Oxford University Press.

Loprieno, Antonio. 1995. *Ancient Egyptian: A Linguistic Introduction*. Cambridge: Cambridge University Press.

Loprieno, Antonio, and Matthias Müller. 2012. Ancient Egyptian and Coptic. In *The Afroasiatic Languages*, edited by Zygmunt Frajzyngier and Erin Shay, 102–144. Cambridge: Cambridge University Press.

Lounsbury, Floyd G. 1946. Stray Number Systems among Certain Indian Tribes. *American Anthropologist* 48 (4): 672–675.

Lowery, George. 1977. Notable Persons in Cherokee History: Sequoyah, or George Gist. *Journal of Cherokee Studies* 2: 385–393.

Lyman, R. Lee, and Judith L. Harpole. 2002. A. L. Kroeber and the Measurement of Time's Arrow and Time's Cycle. *Journal of Anthropological Research* 58 (3): 313–338.

Macfarlane, Alan. 1970. *Witchcraft in Tudor and Stuart England: A Regional and Comparative Study*. London: Routledge and Kegan Paul.

Macfarlane, Alan. 1978. *The Origins of English Individualism: The Family, Property and Social Transition*. Oxford: Basil Blackwell.

Machak, Veronica. 2014. Effects of Typing Awareness Cues on Turn-Taking in Google Chat: Two Methods of Analysis. M.A., Linguistics, Wayne State University.

Mackay, Charles. 1963. *Extraordinary Popular Delusions and the Madness of Crowds*. Wells, VT: Fraser Publishing.

Maher, David W., and John F. Makowski. 2001. Literary Evidence for Roman Arithmetic with Fractions. *Classical Philology* 96 (4): 376–399.

Manca, Vincenzo. 2015. On the Lexicographic Representation of Numbers. *arXiv* preprint *arXiv:1505.00458*.

Mandler, George, and Billie J. Shebo. 1982. Subitizing: An Analysis of Its Component Processes. *Journal of Experimental Psychology: General* 111 (1): 1–22.

Marchand, Trevor. 2018. Toward an Anthropology of Mathematizing. *Interdisciplinary Science Reviews* 43 (3–4): 295–316.

Margetts, Anna. 2007. Learning Verbs without Boots and Straps? The Problem of "Give" in Saliba. In *Cross-Linguistic Perspectives on Argument Structure: Implications for Learnability*, edited by Melissa Bowermann and Penelope Brown, 111–141. Mahwah, NJ: Lawrence Erlbaum.

Marshack, Alexander. 1972. *The Roots of Civilization*. New York: McGraw-Hill.

McDonald, Katherine. 2015. *Oscan in Southern Italy and Sicily: Evaluating Language Contact in a Fragmentary Corpus*. Cambridge: Cambridge University Press.

McPharlin, Paul. 1942. *Roman Numerals, Typographic Leaves and Pointing Hands: Some Notes on Their Origin, History and Contemporary Use*. New York: Typophiles.

Menary, Richard, ed. 2010. *The Extended Mind*. Cambridge, MA: MIT Press.

Mercier, Hugo, and Olivier Morin. 2019. Majority Rules: How Good Are We at Aggregating Convergent Opinions? *Evolutionary Human Sciences* 1: e6. https://doi.org/10.1017/ehs.2019.6.

Miller, George A. 1956. The Magical Number Seven, Plus or Minus Two: Some Limits on Our Capacity for Processing Information. *Psychological Review* 63 (2): 81–97.

Miller, Kevin F., and Jianjun Zhu. 1991. The Trouble with Teens: Accessing the Structure of Number Names. *Journal of Memory and Language* 30 (1): 48–68.

Millet, Nicholas B. 1990. The Narmer Macehead and Related Objects. *Journal of the American Research Center in Egypt* 27: 53–59.

Miner, Horace. 1956. Body Ritual among the Nacirema. *American Anthropologist* 58 (3): 503–507.

Moeller, Korbinian, Samuel Shaki, Silke M. Göbel, and Hans-Christoph Nuerk. 2015. Language Influences Number Processing—A Quadrilingual Study. *Cognition* 136: 150–155.

Montemurro, Marcelo A., and Damián H. Zanette. 2011. Universal Entropy of Word Ordering across Linguistic Families. *PLoS One* 6 (5): e19875.

Montgomery-Anderson, Brad. 2015. *Cherokee Reference Grammar*. Norman: University of Oklahoma Press.

Morley, Iain, and Colin Renfrew, eds. 2010. *The Archaeology of Measurement: Comprehending Heaven, Earth and Time in Ancient Societies*. Cambridge: Cambridge University Press.

Mukhopadhyay, Carol C. 2004. A Feminist Cognitive Anthropology: The Case of Women and Mathematics. *Ethos* 32 (4): 458–492.

Muller, Hermann Joseph, and Lewis Morton Mott-Smith. 1930. Evidence that Natural Radioactivity Is Inadequate to Explain the Frequency of "Natural" Mutations. *Proceedings of the National Academy of Sciences of the United States of America* 16 (4): 277–285.

Murray, Alexander. 1978. *Reason and Society in the Middle Ages*. Oxford: Clarendon Press.

Myers-Scotton, Carol. 1993. *Social Motivations for Codeswitching*. Oxford: Oxford University Press.

Naroll, Raoul, E. C. Benjamin, F. K. Fohl, M. J. Fried, R. E. Hildreth, and J. M. Schaefer. 1971. Creativity: A Cross-Historical Pilot Survey. *Journal of Cross-Cultural Psychology* 2 (2): 181–188.

Naroll, Raoul, Vern L. Bullough, and Frada Naroll. 1974. *Military Deterrence in History: A Pilot Cross-Historical Survey*. Albany: SUNY Press.

Needham, Joseph. 1959. *Science and Civilization in China*, vol. 3: *Mathematics and the Sciences of the Heavens and the Earth*. Cambridge: Cambridge University Press.

Netz, Reviel. 2002. Counter Culture: Towards a History of Greek Numeracy. *History of Science* 40 (3): 321–352.

Nevins, Andrew, David Pesetsky, and Cilene Rodrigues. 2009. Pirahã Exceptionality: A Reassessment. *Language* 85 (2): 355–404.

Newman, Mark E. J. 2005. Power Laws, Pareto Distributions and Zipf's Law. *Contemporary Physics* 46 (5): 323–351.

Nissen, Hans-Jörg, Peter Damerow, and Robert K. Englund. 1993. *Archaic Bookkeeping: Early Writing and Techniques of Economic Administration in the Ancient Near East.* Chicago: University of Chicago Press.

Noël, Marie-Pascale, and Xavier Seron. 1992. Notational Constraints and Number Processing: A Reappraisal of the Gonzalez and Kolers (1982) Study. *Quarterly Journal of Experimental Psychology: Section A* 45 (3): 451–478.

Norman, Donald A. 1988. *The Design of Everyday Things*. New York: Doubleday.

Nothaft, C. Philipp E. 2020. Medieval Europe's Satanic Ciphers: On the Genesis of a Modern Myth. *British Journal for the History of Mathematics* 35 (2): 107–136.

Nougayrol, Jean. 1972. Notes brèves 12. *Revue d'assyriologie et d'archéologie orientale* 66 (1): 93–96.

Nykl, Alois R. 1926. The Quinary-Vigesimal System of Counting in Europe, Asia, and America. *Language* 2 (3): 165–173.

Olson, David R. 1994. *The World on Paper: The Conceptual and Cognitive Implications of Writing and Reading*. Cambridge: Cambridge University Press.

O'Neil, Cathy. 2016. *Weapons of Math Destruction: How Big Data Increases Inequality and Threatens Democracy*. New York: Crown.

Otis, Jessica. 2017. "Set Them to the Cyphering Schoole": Reading, Writing, and Arithmetical Education, circa 1540–1700. *Journal of British Studies* 56 (3): 453–482.

Ouyang, Xiaoli. 2016. The Mixture of Sexagesimal Place Value and Metrological Notations on the Ur III Girsu Tablet BM 19027. *Journal of Near Eastern Studies* 75 (1): 23–41.

Overmann, Karenleigh A. 2015. Numerosity Structures the Expression of Quantity in Lexical Numbers and Grammatical Number. *Current Anthropology* 56 (5): 638–653.

Overmann, Karenleigh A. 2016. Beyond Writing: The Development of Literacy in the Ancient Near East. *Cambridge Archaeological Journal* 26 (2): 285–303.

Paulos, John Allen. 1988. *Innumeracy: Mathematical Illiteracy and Its Consequences*. New York: Macmillan.

Pechenick, Eitan Adam, Christopher M. Danforth, and Peter Sheridan Dodds. 2015. Characterizing the Google Books Corpus: Strong Limits to Inferences of Socio-cultural and Linguistic Evolution. *PLoS One* 10 (10): e0137041.

Perdue, Theda, and Michael D. Green. 2007. *The Cherokee Nation and the Trail of Tears*. New York: Penguin.

Peregrine, Peter N. 2001. Cross-Cultural Comparative Approaches in Archaeology. *Annual Review of Anthropology* 30 (1): 1–18.

Peregrine, Peter N. 2004. Cross-Cultural Approaches in Archaeology: Comparative Ethnology, Comparative Archaeology, and Archaeoethnology. *Journal of Archaeological Research* 12 (3): 281–309.

Pericliev, Vladimir. 2004. Universals, Their Violation and the Notion of Phonologically Peculiar Languages. *Journal of Universal Language* 5 (1): 85–117.

Periton, Cheryl. 2015. The Medieval Counting Table Revisited: A Brief Introduction and Description of Its Use during the Early Modern Period. *BSHM Bulletin: Journal of the British Society for the History of Mathematics* 30 (1): 35–49.

Periton, Cheryl. 2017. Education, the Development of Numeracy and Dissemination of Hindu-Arabic Numerals in Early Modern Kent. PhD diss., Canterbury Christ Church University.

Pettinato, Giovanni. 1981. *The Archives of Ebla: An Empire Inscribed in Clay*. Garden City, NY: Doubleday.

Pihan, Antoine Paulin. 1860. *Exposé des signes de numération usités chez les peuples orientaux anciens et modernes*. Paris: Imprimerie impériale.

Pinnow, Heinz-Jürgen. 1972. Schrift und Sprache in den Werken Lako Bodras im Gebiet der Ho von Singbhum (Bihar). *Anthropos* 67 (5–6): 822–857.

Plofker, Kim. 2009. *Mathematics in India*. Princeton: Princeton University Press.

Pohl, Frederik. 1966. *Digits and Dastards*. New York: Ballantine Books.

Pollock, Sheldon. 2009. Future Philology? The Fate of a Soft Science in a Hard World. *Critical Inquiry* 35 (4): 931–961.

Pompeo, Flavia. 2015. I greci a Persepoli. Alcune riflessioni sociolinguistiche sulle iscrizioni greche nel mondo iranico. In *Contatto interlinguistico fra presente e passato*, edited by Carlo Consani, 149–172. Milan: LED.

Pospisil, Leopold, and Derek J. de Solla Price. 1966. A Survival of Babylonian Arithmetic in New Guinea? *Indian Journal of the History of Science* 1: 30–33.

Pospisil, Leopold, and Derek J. de Solla Price. 1976. Reckoning and Racism. *Journal of the Polynesian Society* 85 (3): 382–383.

Postgate, Nicholas, Tao Wang, and Toby Wilkinson. 1995. The Evidence for Early Writing: Utilitarian or Ceremonial? *Antiquity* 69 (264): 459–480.

Pott, August Friedrich. 1847. *Die quinare und vigesimale Zählmethode bei Völkern aller Welttheile: Nebst ausführlicheren Bemerkungen über die Zahlwörter indogermanischen Stammes und einem Anhange über Fingernamen*. Halle: C. A. Schwetschke und Sohn.

Powell, Marvin A. 1972. The Origin of the Sexagesimal System: The Interaction of Language and Writing. *Visible Language* 6 (1): 5–18.

Preston, Jean F. 1994. Playing with Numbers: Some Mixed Counting Methods Found in French Medieval Manuscripts at Princeton. In *Medieval Codicology, Iconography, Literature and Translation: Studies for Keith Val Sinclair*, edited by Peter R. Monks and D. D. R. Owen, 74–78. Leiden: Brill.

Prince, Alan, and Paul Smolensky. 1993. *Optimality Theory: Constraint Interaction in Generative Grammar.* New Brunswick: Rutgers University Center for Cognitive Science.

Pritchard, Violet. 1967. *English Medieval Graffiti.* Cambridge: Cambridge University Press.

Proust, Christine. 2009. Numerical and Metrological Graphemes: From Cuneiform to Transliteration. *Cuneiform Digital Library Journal* 1. http://www.cdli.ucla.edu/pubs/cdlj/2009/cdlj2009_001.html.

Quibell, James Edward, Frederick Wastie Green, and William Matthew Flinders Petrie. 1900. *Hierakonpolis*. Vol. 4. London: B. Quaritch.

Reboul, Anne. 2012. Language: Between Cognition, Communication and Culture. *Pragmatics and Cognition* 20 (2): 295–316.

Rendell, Luke, Robert Boyd, Daniel Cownden, Marquist Enquist, Kimmo Eriksson, Marc W. Feldman, Laurel Fogarty, Stefano Ghirlanda, Timothy Lillicrap, and Kevin N. Laland. 2010. Why Copy Others? Insights from the Social Learning Strategies Tournament. *Science* 328 (5975): 208–213.

Renfrew, Colin. 1990. *Archaeology and Language: The Puzzle of Indo-European Origins.* Cambridge: Cambridge University Press.

Richards, Anne. 1996. Does Clock-Watching Make You Clockwise? *Memory* 4 (1): 49–58.

Richerson, Peter J., and Robert Boyd. 2005. *Not by Genes Alone: How Culture Transformed Human Evolution.* Chicago: University of Chicago Press.

Rips, Lance J. 2011. *Lines of Thought: Central Concepts in Cognitive Psychology.* New York: Oxford University Press.

Roberts, Gareth. 2000. Bilingualism and Number in Wales. *International Journal of Bilingual Education and Bilingualism* 3 (1): 44–56.

Rogers, Everett M. 1962. *Diffusion of Innovations.* New York: Free Press.

Rojo-Garibaldi, Berenice, Costanza Rangoni, Diego L. González, and Julyan H. E. Cartwright. 2020. Non-Power Positional Number Representation Systems, Bijective Numeration, and the Mesoamerican Discovery of Zero. *arXiv* preprint *arXiv:2005.10207.*

Rosch, Eleanor H. 1973. Natural Categories. *Cognitive Psychology* 4 (3): 328–350.

Rouse, Richard H., and Mary A. Rouse. 1979. *Preachers, Florilegia and Sermons: Studies on the Manipulus florum of Thomas of Ireland.* Toronto: Pontifical Institute of Medieval Studies.

Rymer, Thomas. 1740. *Foedera, conventiones, literae et cujuscunque generis acta publica, inter reges Angliae et alios quosvis imperatores, reges, pontifices, principes, vel communitates, ab ineunte saeculo duodecimo, viz. ab anno 1101. ad nostra usque tempora, habita aut tractata: ex autographis, infra secretiores archivorum regiorum thesaurarias per multa saecula reconditis, fideliter exscripta*. The Hague: Joannes Neaulme.

Saenger, Paul. 1997. *Space between Words: The Origins of Silent Reading*. Stanford: Stanford University Press.

Saidan, Ahmad S. 1966. The Earliest Extant Arabic Arithmetic: *Kitāb al-Fuṣūl fī al Ḥisāb al-Hindī* of Abū al-Ḥasan, Aḥmad ibn Ibrāhīm al-Uqlīdisī. *Isis* 57 (4): 475–490.

Sapir, Edward. 1921. *Language: An Introduction to the Study of Speech*. New York: Harcourt, Brace.

Saxe, Geoffrey B. 1981. Body Parts as Numerals: A Developmental Analysis of Numeration among the Oksapmin in Papua New Guinea. *Child Development* 52 (1): 306–316.

Saxe, Geoffrey B. 1982. Developing Forms of Arithmetical Thought among the Oksapmin of Papua New Guinea. *Developmental Psychology* 18 (4): 583–594.

Saxe, Geoffrey B. 2012. *Cultural Development of Mathematical Ideas: Papua New Guinea Studies*. New York: Cambridge University Press.

Sayce, Richard Anthony. 1966. Compositorial Practices and the Localization of Printed Books, 1530–1800. *The Library* 5 (1): 1–45.

Schärlig, Alain. 2004. Un bas-relief à Trèves: Ces romains calculent, ils ne jouent pas! *Antike Kunst* 47: 65–71.

Schlimm, Dirk, and Hansjörg Neth. 2008. Modeling Ancient and Modern Arithmetic Practices: Addition and Multiplication with Arabic and Roman Numerals. *CogSci 2008*, 2097–2102. Red Hook, NY: Curran.

Schmandt-Besserat, Denise. 1992. *Before Writing*, vol. 1: *From Counting to Cuneiform*. Austin: University of Texas Press.

Schmitt, Rüdiger. 1989. Ein altiranisches Flüssigkeitsmass, *mariš. In *Indogermanica Europaea, Festschrift für Wolfgang Meid*, edited by Karin Heller, Oswald Panagl, and Johann Tischler, 301–315. Graz: University of Graz.

Schub, Pincus. 1932. A Mathematical Text by Mordecai Comtino (Constantinople, XV Century). *Isis* 17 (1): 54–70.

Scott, James C. 2017. *Against the Grain: A Deep History of the Earliest States*. New Haven: Yale University Press.

Scribner, Sylvia, and Michael Cole. 1981. *The Psychology of Literacy*. Cambridge, MA: Harvard University Press.

Seaver, Nick. 2018. What Should an Anthropology of Algorithms Do? *Cultural Anthropology* 33 (3): 375–385.

Segal, Daniel Alan, and Sylvia Junko Yanagisako. 2005. *Unwrapping the Sacred Bundle: Reflections on the Disciplining of Anthropology*. Durham: Duke University Press.

Seiler, Hansjakob. 1995. Iconicity between Indicativity and Predicativity. In *Iconicity in Language*, edited by Raffaele Simone, 141–151. Amsterdam: John Benjamins.

Serafini, Luigi. 1983. *Codex Seraphinianus*. New York: Abbeville Press.

Shennan, Stephen. 2002. *Genes, Memes and Human History: Darwinian Archaeology and Cultural Evolution*. London: Thames and Hudson.

Shipley, Frederick W. 1902. Numeral Corruptions in a Ninth Century Manuscript of Livy. *Transactions and Proceedings of the American Philological Association* 33: 45–54.

Sigurd, Bengt. 1988. Round Numbers. *Language in Society* 17 (2): 243–252.

Small, Jocelyn Penny. 1997. *Wax Tablets of the Mind: Cognitive Studies of Memory and Literacy in Classical Antiquity*. London: Routledge.

Smullyan, Raymond M. 1961. *Theory of Formal Systems*. Princeton: Princeton University Press.

Spengler, Oswald. 1926. *The Decline of the West*. London: G. Allen and Unwin.

Stauder, Andréas. 2018. On System-Internal and Differential Iconicity in Egyptian Hieroglyphic Writing. *Signata. Annales des sémiotiques/Annals of Semiotics* 9: 365–390.

Stedall, Jacqueline A. 2001. Of Our Own Nation: John Wallis's Account of Mathematical Learning in Medieval England. *Historia Mathematica* 28 (2): 73–122.

Stigler, James W. 1984. "Mental Abacus": The Effect of Abacus Training on Chinese Children's Mental Calculation. *Cognitive Psychology* 16 (2): 145–176.

Stolper, Matthew W., and Jan Tavernier. 2007. From the Persepolis Fortification Archive Project, 1: An Old Persian Administrative Tablet from the Persepolis Fortification. *ARTA: Achaemenid Research on Texts and Archaeology* 1. http://hdl.handle.net/2078/84230.

Stone, Lawrence. 1949. Elizabethan Overseas Trade. *Economic History Review* 2 (1): 30–58.

Struik, Dirk. 1968. The Prohibition of the Use of Arabic Numerals in Florence. *Archives internationales d'histoire des sciences* 21: 291–294.

Surowiecki, James. 2004. *The Wisdom of Crowds*. New York: Doubleday.

Swadesh, Morris. 1936. Review: *The Psycho-biology of Language: An Introduction to Dynamic Philology*. *American Anthropologist* 38 (3): 505–506.

Tagliamonte, Sali A. 2016. So Sick or So Cool? The Language of Youth on the Internet. *Language in Society* 45 (1): 1–32.

Tagliavini, Carlo. 1949. Di alcune denominazioni della papilla. *Annali dell'Istituto Universitario di Napoli*, n.s. 3: 341–378.

Taisbak, Christian M. 1965. Roman Numerals and the Abacus. *Classica et Medievalia* 26: 147–160.

Thompson, David. 2008. *Watches*. London: British Museum.

Threatte, Leslie. 1980. *The Grammar of Attic Inscriptions: Phonology*. Vol. 1. Berlin: De Gruyter.

Tod, Marcus N. 1979. *Ancient Greek Numerical Systems: Six Studies*. Chicago: Ares Publishers.

Toynbee, Arnold Joseph. 1934–1961. *A Study of History*. 12 vols. Oxford: Oxford University Press.

Trigger, Bruce G. 1980. Archaeology and the Image of the American Indian. *American Antiquity* 45 (4): 662–676.

Trigger, Bruce G. 1990. Monumental Architecture: A Thermodynamic Explanation of Symbolic Behaviour. *World Archaeology* 22 (2): 119–132.

Trigger, Bruce G. 1991. Distinguished Lecture in Archeology: Constraint and Freedom—A New Synthesis for Archeological Explanation. *American Anthropologist* 93 (3): 551–569.

Trigger, Bruce G. 1998. Writing Systems: A Case Study in Cultural Evolution. *Norwegian Archaeological Review* 31 (1): 39–62.

Tylor, Edward B. 1871. *Primitive Culture: Researches into the Development of Mythology, Philosophy, Religion, Art, and Custom*. London: J. Murray.

University of Chicago Press. 1906. *A Manual of Style, Containing Typographical and Other Rules for Authors, Printers, and Publishers Recommended by the University of Chicago Press, Together with Specimens of Type*. Chicago: University of Chicago Press.

Urban, Greg. 2010. A Method for Measuring the Motion of Culture. *American Anthropologist* 112 (1): 122–139.

Valenti, Gianluca. 2009. The Magical Number Seven and the Early Romance Poetry. *Cognitive Philology* 2. https://annalidibotanica.uniroma1.it/index.php/cogphil/article/view/8812.

Van Stone, Mark. 2011. It's Not the End of the World: What the Ancient Maya Tell Us about 2012. *Archaeoastronomy* 24: 12–36.

Veblen, Thorstein. 1899. *The Theory of the Leisure Class*. New York: A. M. Kelley.

Velserus, Marcus. 1682. *Opera historica et philologica, sacra et profana: In quibus Historia Boica, Res Augustanae, Conversio & Passio SS. martyrum, Afrae, Hilariae … continentur.*

Accessit P. Optatiani Porphyrii Panegyricus ... nec non vita, genius, et mors auctoris ... Accurante Christophoro Arnoldo. Nuremberg: Typis ac sumtibus Wolfgangi Mauritii & Filiorum Johannis Andreae.

Vernus, Pascal. 2004. Le syntagme de quantification en égyptien de la première phase: Sur les relations entre textes des pyramides et textes des sarcophages. In *D'un monde à l'autre: Textes des pyramides & textes des sarcophages*, edited by Susanne Bickel and Bernard Mathieu, 279–311. Cairo: Institut Français d'Archéologie Orientale.

Walker, Willard, and James Sarbaugh. 1993. The Early History of the Cherokee Syllabary. *Ethnohistory* 40 (1): 70–94.

Wallerstein, Immanuel. 1974. *The Modern World-System*, vol. 1: *Capitalist Agriculture and the Origins of European World-Economy in the Sixteenth Century*. New York: Academic Press.

Wallis, John. 1685. *A Treatise of Algebra, Both Historical and Practical*. London: John Playford.

Wardley, Peter, and Pauline White. 2003. The Arithmeticke Project: A Collaborative Research Study of the Diffusion of Hindu-Arabic Numerals. *Family and Community History* 6 (1): 5–17.

Watson, Richard A. 1976. Inference in Archaeology. *American Antiquity* 41 (1): 58–66.

Whatmough, Joshua. 1952. Natural Selection in Language. *Scientific American* 186 (4): 82–87.

Wiese, Heike. 2003. *Numbers, Language, and the Human Mind*. Cambridge: Cambridge University Press.

Williams, Burma P., and Richard S. Williams. 1995. Finger Numbers in the Greco-Roman World and the Early Middle Ages. *Isis* 86 (4): 587–608.

Williams, Jack. 1997. Numerals and Numbering in Early Printed English Bibles and Associated Literature. *Journal of the Printing Historical Society* 26: 5–13.

Witthoft, Nathan, and Jonathan Winawer. 2006. Synesthetic Colors Determined by Having Colored Refrigerator Magnets in Childhood. *Cortex* 42 (2): 175–183.

Witthoft, Nathan, and Jonathan Winawer. 2013. Learning, Memory, and Synesthesia. *Psychological Science* 24 (3): 258–265.

Wobst, H. Martin. 1978. The Archaeo-ethnology of Hunter-Gatherers or the Tyranny of the Ethnographic Record in Archaeology. *American Antiquity* 43 (2): 303–309.

Wohlgemuth, Jan, and Michael Cysouw, eds. 2010. *Rethinking Universals: How Rarities Affect Linguistic Theory*. Berlin: De Gruyter.

Wolf, Eric R. 1982. *Europe and the People without History*. Berkeley: University of California Press.

Woods, Christopher. 2017. The Abacus in Mesopotamia: Considerations from a Comparative Perspective. In *The First Ninety Years: A Sumerian Celebration in Honor of Miguel Civil*, edited by Luis Feliu, Fumi Karahashi, and Gonzalo Rubio, 416–478. Berlin: De Gruyter.

Woolard, Kathryn A., and Bambi B. Schieffelin. 1994. Language Ideology. *Annual Review of Anthropology* 23 (1): 55–82.

Worthen, Shana Sandlin. 2006. The Memory of Medieval Inventions, 1200–1600: Windmills, Spectacles, Mechanical Clocks, and Sandglasses. PhD diss., University of Toronto.

Yellowhorn, Eldon. 2006. Understanding Antiquity: Bruce Trigger on His Life's Work in Archaeology—An Interview. *Journal of Social Archaeology* 6 (3): 307–327.

Yong, Bhikhshu Jin. 2008. How Large Is One Asamkhyeya? *Vajra Bodhi Sea* 462 (November 2008): 42–44.

Young, Dwight W. 1988. A Mathematical Approach to Certain Dynastic Spans in the Sumerian King List. *Journal of Near Eastern Studies* 47 (2): 123–129.

Zhang, Jiajie, and Donald A. Norman. 1995. A Representational Analysis of Numeration Systems. *Cognition* 57 (3): 271–295.

Zide, Norman. 1996. Scripts for Munda Languages. In *The World's Writing Systems*, edited by Peter T. Daniels and William Bright, 612–618. New York: Oxford University Press.

Zipf, George Kingsley. 1929. Relative Frequency as a Determinant of Phonetic Change. *Harvard Studies in Classical Philology* 40: 1–95.

Zipf, George Kingsley. 1935. *The Psycho-biology of Language: An Introduction to Dynamic Philology*. Boston: Houghton Mifflin.

Zipf, George Kingsley. 1949. *Human Behavior and the Principle of Least Effort: An Introduction to Human Ecology*. Cambridge, MA: Addison-Wesley.

Index